DFG

Analyses of Hazardous Substances in Air

WILEY-VCH

DFG Deutsche Forschungsgemeinschaft

Analyses of Hazardous Substances in Air

Volume 6

edited by Antonius Kettrup
Working Group Analytical Chemistry

Commission for the Investigation of Health Hazards of Chemical Compounds in the Work Area
(Chairman: Helmut Greim)

WILEY-VCH

Prof. Dr. med. Helmut Greim
Senatskommission
zur Prüfung gesundheitsschädlicher Arbeitsstoffe
der Deutschen Forschungsgemeinschaft
Technische Universität München
Hohenbachernstr. 15–17
D-85354 Freising-Weihenstephan

Prof. Dr. rer. nat. Dr. h.c. Antonius Kettrup
GSF-Forschungszentrum für Umwelt und Gesundheit
Institut für Ökologische Chemie
Ingolstädter Landstraße 1
D-85764 Neuherberg

Translators: Julia Handwerker-Sharman,
Dr. Karl-Heinz Ohrbach, Dr. Ann E. Wild

Deutsche Bibliothek Cataloguing-in-Publication Data:

Analyses of hazardous substances in air / DFG, Dt. Forschungsgemeinschaft; Comm. for the Investigation of Health Hazards of Chem. Compounds in the Work Area. – Weinheim ; New York ; Chichester ; Brisbane ; Singapore ; Toronto : Wiley-VCH
Erscheint unregelmäßig. – Aufnahme nach Vol. 1 (1991)
Vol. 1 (1991) –

Printed on acid-free paper.

Composition: ProSatz Unger, Weinheim.
Printing: Strauss Offsetdruck GmbH, Mörlenbach
Bookbinding: J. Schäffer GmbH & Co. KG, Grünstadt

Printed in the Federal Republic of Germany.

Preface

This is the sixth volume in the series of the English translations of analytical methods for the determination of hazardous chemicals in workplace air by the Working Group "Analytical Chemistry" of the Commission for the Investigation of Health Hazards of Chemical Compounds in the Work Area of the Deutsche Forschungsgemeinschaft.

The Working Group "Analytical Chemistry" is continuing its efforts to elaborate, validate and introduce suitable methods for determining hazardous chemicals in workplace air. At present, more than 140 methods are published in the German edition of Analyses of Hazardous Substances in Air. Since Volumes 1 to 5 of the English edition aroused considerable interest, the Working Group has provided another series of 14 standardised methods in this volume. Volume 7 is already in preparation.

The speciality of this volume is the fact that methods for real solvents, which are used in the working area, are developed. On the other hand the theoretical background of passive sampling and analytical quality control is for the first time demonstrated.

I wish to acknowledge the careful work of the members of the Working Group "Analytical Chemistry" and the contribution made by the authors and examiners of these methods, as well as the accurate work of the translators of this volume. My special thanks go in particular to M.R. Lahaniatis and A. Kettrup for their successful work and great personal engagement.

On behalf of the Commission, I am pleased to express the hope that this publication will be beneficial to the protection of the health and improve the working conditions of those exposed.

H. Greim
Chairman of the Commission for the
Investigation of Health Hazards of Chemical
Compounds in the Work Area

Foreword

Health protection, the goal of the Occupational Exposure Limits (OEL, i.e. the maximum permissible concentrations in the workplace atmosphere), can be monitored by systematic and quantitative determination of the chemical compounds used at the work place.
The concentrations of chemical compounds in the air of work areas fluctuate around the level of a statistical mean value. However, concentrations far above (or below) the mean value can occur for brief periods. These differences must be accounted for in the analytical procedure. As far as possible, a representative measure of the concentration of a compound and its fluctuation should be made. Therefore, sampling should be conducted continuously during a complete work period or by representative random sampling, the work pattern must also be taken into consideration. The measurements should be scheduled to yield a value for the mean concentration, or a statistically significant estimate of this, within a particular examination period. Whenever possible, sampling should be carried out in the breathing zone in close proximity to the employee.
Research and development of methods suitable for routine use is currently under way in the Working Subgroup "Analyses of Hazardous Substances in Air of Work Areas" in line with the objectives of the Working Group "Analytical Chemistry" of the Commission for the Investigation of Health Hazards of Chemical Compounds in the Work Area of the Deutsche Forschungsgemeinschaft.
This Working Subgroup has compiled and published established methods for the analysis of air in German. In response to the worldwide demand for methods for the analysis of chemicals in air, the Commission for the Investigation of Health Hazards of Chemical Compounds in the Work Area has decided to make the German contributions in this field available to an international audience by publishing an English edition.
Volume 6 comprises a further 7 analytical methods for determining hazardous compounds. In addition the volume contains 7 methods on solvent mixtures, which are very often present at different work areas containing substances of different classes.
In the meantime for the special group of volatiles from "Ottokraftstoff" (fuel) a limit value is established (TRGS 900/901).
Furthermore, this volume presents two general chapters on passive sampling and on analytical quality assurance. The chapter on passive sampling describes fundamental principles and the influences on the sampling procedure.
We like to thank the members and guests of the Working Subgroup without whose voluntary services this collection of methods would not have been possible. We thank the Deutsche Forschungsgemeinschaft for financial and organisational help in the development of our activities.

Our thanks go also to our publisher Dr. Eva E. Wille of the Wiley-VCH Verlag with whom we have enjoyed long-standing and efficient collaboration. We also wish to thank Mrs. Julia Handwerker-Sharman for translation.

J. Angerer
Chairman of the Working Group "Analytical Chemistry" of the Commission for the Investigation of Health Hazards of Chemical Compounds in the Work Area

A. Kettrup
Chairman of the Working Subgroup "Analyses of Hazardous Substances in Air of Work Areas"

Contents

Working Group "Analytical Chemistry" of the Commission of the Deutsche Forschungsgemeinschaft for the Investigation of Health Hazards of Chemical Compounds in the Work Area

Organization

The Working Group "Analytical Chemistry" was established in 1969. Under the chairmanship of Prof. Dr. J. Angerer at the present it includes two Working Subgroups:

"Air Analyses"
(Leader: Prof. Dr. rer. nat. Dr. h.c. A. Kettrup)

"Analyses of Hazardous Substances in Biological Materials"
(Leaders: Prof. Dr. J. Angerer and Chem.-Ing. K.H. Schaller).

The participants, who have been invited to collaborate on a Working Subgroup by the leaders, are experts in the field of technical and medical protection against chemical hazards at the workplace. A list of members and guests of "Analyses of Hazardous Substances in Air" is given at the end of this volume.

Objectives and operational procedure

The two analytical subgroups are charged with the task of preparing methods for the determination of hazardous industrial materials in the air of the workplace or to determine these hazardous materials or their metabolic products in biological specimens from the persons working there. Within the framework of the existing laws and regulations, these analytical methods are useful for ambient monitoring at the workplace and biological monitoring of the exposed persons.
In addition to the working out the analytical procedure, these subgroups are concerned with the problems of the preanalytical phase (specimen collection, storage, transport), the statistical quality control, as well as the interpretation of the results.

Development, examination, release, and quality of the analytical methods

In its selection of suitable analytical methods, the Working Group is guided mainly by the relevant scientific literature and the expertise of the members and guests of the Working Subgroup. If appropriate analytical methods are not available they are worked out within the Working Group. The leader designates an author, who assumes the task of developing and formulating a method proposal. The proposal is examined experimentally by at least one other member of the project, who then submits a written report of the results of the examination. As a matter of principle the examination must encompass all phases of the proposed analytical procedure. The examined method is then laid before the members of the subgroups for consideration. After hearing the judgement of the author and the examiner they can approve the method. The method can then be re-

leased for publication after a final meeting of the leader of the Working Group "Analytical Chemistry" with the subgroup leaders, authors, and examiners of the method.
Under special circumstances an examined method can released for publication by the leader of the Working Group after consultation with the subgroup leaders.
Only methods for which criteria of analytical reliability can be explicitly assigned are released for publication. The values for inaccuracy, imprecision, detection limits, sensitivity, and specificity must fulfil the requirements of statistical quality control as well as the specific standards set by occupational health. The above procedure it meant to guarantee that only reliably functioning methods are published, which are not only reproducible within the framework of the given reliability criteria in different laboratories, but also can be monitored over the course of time.
In the selection and development of a method for determining a particular substance the Working Group has given the analytical reliability of the method precedence over aspects of simplicity and economy.

Publications of the working group

Methods released by the Working Group are published in the Federal Republic of Germany, by the Deutsche Forschungsgemeinschaft as a loose-leaf collection entitled "Analytische Methoden zur Prüfung gesundheitsschädlicher Arbeitsstoffe" (WILEY-VCH Verlag, Weinheim, FRG). The collection at present consists of two volumes:

Volume I "Luftanalysen"
Volume II "Analysen in biologischem Material".

These methods are also to be published in an English edition. Volume 1 to 7 of "Analyses of Hazardous Substances in Biological Materials" have already been published. The work at hand represents the sixth English issue of "Analyses of Hazardous Substances in Air".

Withdrawal of methods

An analytical method that is made obsolete by new developments or discoveries in the fields of instrumental analysis or occupational health and toxicology can be replaced by a more efficient method. After consultation with the membership of the relevant project and with the consent of the leader of the Working Group, the subgroup leader is empowered to withdraw the old method.

Preliminary Remarks

Passive sampling

Contents

1 Passive sampling

1.1 The fundamental principles of passive sampling

1.1.1 The use of diffusion for passive sampling

Passive sampling makes use of the laws of diffusion. There are various sampling systems for passive sampling.
Such systems are made up of a sampling phase (e. g. activated carbon) mounted a short distance behind an opening of known cross-section which is in contact with the ambient air. The molecules are transported from the opening to the collection phase by diffusion. To minimise the effect of air currents, either the whole diffusion layer can be made of porous material or the opening of the diffusion sampler can be covered with a porous membrane. In the latter case the substance is transported in a first step by permeation, followed by diffusion [1]. With very small diffusion cross-sections and very long diffusion paths, such layers or membranes are not necessary.
Figure 1 shows diagrams of diffusion samplers:

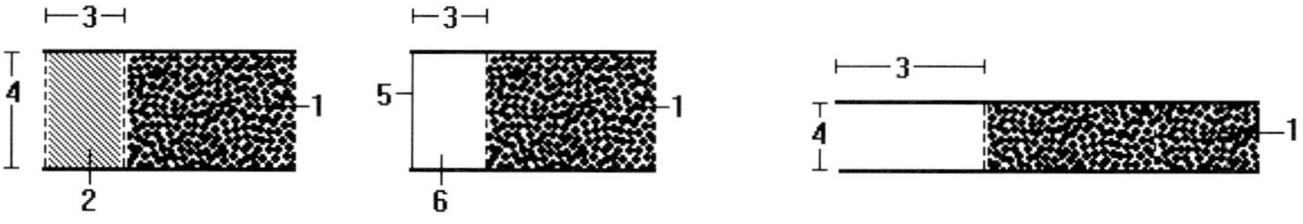

Fig. 1. Diagrams of diffusion samplers. 1 sampling phase, 2 covering (porous material), 3 diffusion path, 4 diffusion cross-section, 5 porous membrane, 6 air volume (diffusion chamber).

Diffusion samplers used to determine the concentration of a substance directly in accordance with the standard DIN EN 838, so-called reaction samplers, are called type A [2].
Characteristic of reaction samplers is that the substance to be collected produces a specific chemical reaction with a reagent on the surface of a collection phase; it makes sense that this reaction should result in a colour change. The concentration of the substance to be collected is determined either by estimating the intensity of the colour, which is approximately proportional to the amount of substance reacted, or by using a scale for the length of the coloured zone.
On the other hand, diffusion samplers used for indirect determination of the concentration with sampling by diffusion and adsorption and subsequent analysis in a separate step are called type B samplers.

1.1.2 The laws of diffusion

Diffusion is the molecular transport of a substance and is observed in gaseous, liquid and solid media. The forces behind this substance transport are differences in concentration, partial pressure and temperature. Characteristic of diffusion processes is that transport of the substance takes place without any apparent flow, i. e. without movement of the whole medium.

In general, diffusion is understood to be the process whereby molecules migrate within a system as a result of differences in potential [3].

During passive sampling the only force behind the transport of the substance is the difference in the concentration of the substance in the ambient air and that on the surface of the collection phase. To calculate the amount of substance which has diffused within a certain period of time, geometric parameters, such as the length of the diffusion path over which the concentration differs, and the cross-section through which a certain amount of substance diffuses within a defined period of time, must be taken into account. The diffusion process can be described mathematically by Fick's first law of diffusion [4]:

$$\frac{\Delta n}{\Delta t} = -D \cdot q \cdot \frac{\Delta c}{\Delta x} \tag{1}$$

where:

n	is the number of moles of the substance transported
$\Delta n/\Delta t$	is the rate of diffusion in mole/s
c	is the substance concentration in mole/cm^3
x	is the diffusion path in cm
$\Delta c/\Delta x$	is the difference in concentration in mole/cm^4
q	is the diffusion cross-section in cm^2
t	is the time in s
D	is the diffusion coefficient in cm^2/s

Assuming that, as everything is adsorbed, the concentration of the substance directly above the collection phase is zero and that the concentration of the substance in the vicinity of the sampler corresponds with that in the ambient air, formula (1) can be simplified [1]:

$$n = -D \cdot q \cdot \frac{c}{x} \cdot t \tag{2}$$

The diffusion coefficient is a substance-specific proportionality factor with the dimensions cm^2/s, which is dependent on pressure and temperature and other parameters. Its value can be read from tables or determined experimentally. The following equation can be used to correct for pressure and temperature [5]:

$$D_{\mathrm{T}} = D_{298} \cdot \left(\frac{T}{298}\right)^{1.5} \frac{101.3}{p} \tag{3}$$

where:
D_T is the corrected diffusion coefficient at T and p in cm^2/s
D_{298} is the diffusion coefficient at 298 K and 101.3 kPa in cm^2/s
T is the actual temperature in K
p is the actual air pressure in kPa

The distance moved by the substance per unit time and concentration in the ambient air can be defined as the adsorption or diffusion rate. It can be calculated from Fick's first law of diffusion (Equation 2):

$$\frac{n}{(c \cdot t)} = D \cdot \frac{q}{x} \tag{4}$$

Correspondingly the rate of adsorption U can be calculated from

$$U = \frac{m}{(c \cdot t)} \tag{5}$$

where:
m is the weight of substance adsorbed in mg.

For comparison of passive with active sampling, the rate of adsorption is also commonly given in mL/s or mL/min, i.e. it is expressed as the weight of the substance present in the given volume per unit time. It must be remembered, however, that the volume flowrate is fictive. In reality, transport of the substance takes place without any airflow. Furthermore, Equation (4) shows that the rate of adsorption is dependent solely on the geometry of the diffusion sampler and, of course, on the substance-specific diffusion coefficient. A larger diffusion cross-section and a shorter diffusion path therefore yields a higher rate of adsorption, while a reversal in the relationship between these dimensions results in less of the substance being transported per unit time and concentration.

1.1.3 The effect of environmental conditions on passive sampling

1.1.3.1 The effect of fluctuations in concentration

If a diffusion sampler is exposed to a constant concentration of a substance, after a brief equilibration period, a linear concentration gradient is established in the diffusion zone of the sampler and determines the rate of adsorption. The time t_{equil} needed for equilibration of the linear concentration gradient may be calculated from the following equation:

$$t_{equil} = 1.5 \cdot X^2 D \tag{6}$$

If the concentration fluctuates, the rate of adsorption should adjust as quickly as possible to these changes. The time needed is referred to as the response time. If the concentration gradient is linear, the response time can also be defined as the average period t_r spent by the individual molecules in the diffusion zone before adsorption [6].

$$t_{\mathrm{r}} = \frac{\left(\frac{c \cdot q}{2x}\right)}{\left(\frac{D \cdot q \cdot c}{x}\right)} = \frac{x^2}{2D} \tag{7}$$

It may be readily seen from Equation (7) that samplers with short diffusion paths, which according to Equation (4) also have a high rate of adsorption, have a very short response time and are therefore well-suited for use with fluctuating concentrations. Generally the response times of diffusion samplers are in the range from 0.5 seconds to several seconds [7].

1.1.3.2 The effect of temperature and air pressure

Temperature and air pressure influence the diffusion coefficient and sampling rate and therefore also the amount adsorbed. According to Equation (3), this influence is small and in practice can be ignored. The theoretical increase in the adsorption rate is 0.2%/K [8].
For reaction samplers (type A), the manufacturers generally prescribe a temperature range from about 0 °C to 40 °C.

1.1.3.3 The effect of humidity

With diffusion samplers (type B), whose sampling phase is able to adsorb water, interference from humidity must be expected.
A general statement cannot be made, as substance-specific properties and also the geometry of the sampler and the type and capacity of the sampling phase play a role.
For example, Ikeda *et al.* [9] investigated the behaviour of activated carbon as sampling phase for *n*-hexane, ethyl acetate and toluene at relative humidities between 40% and 93%. Relative humidity above 40% was found to have considerable influence on the sampling of *n*-hexane but not on that of ethyl acetate or toluene. Many authors regard the threshold for the influence of water vapour to be around 50% relative humidity [10]. *Blome* und *Hennig* [11] report for various types of sampler an influence of relative humidity on the rate of adsorption of toluene, 1-butyl acetate, 2-butanone and dichloromethane in the range from 10–75%. It was found that samplers with a high weight of an adsorbent such as activated carbon, and therefore a high adsorbing capacity, are advantageous.
With reaction samplers the analytical principle is such that the substance to be sampled is not expected to be displaced. However, the effect of humidity on such samplers depends on the specificity of the reagent. The manufacturer's instructions should be observed; these usually state a maximum water content of about 15 mg/litre (which corresponds to 65% relative humidity at 25 °C).

To conclude, the effect of water vapour or humidity must be investigated and known for each analysis to be carried out.

1.1.3.4 The effect of air currents

Air currents can make the diffusion rate different from that in still air. Fick's first law of diffusion is only valid for constant conditions, i.e. the concentration of the substance in the environment of the sampler must be constant. This ideal situation is not always found in practice. Convection currents, laminar currents and still layers of air in the area of the sampler can cause interference, so that quantification of the amount of substance adsorbed using Fick's law of diffusion can lead to incorrect results.

If a diffusion sampler is exposed in still air, the concentration of the substances being sampled is reduced in the immediate vicinity of the sampler as a result of diffusion to the surface of the adsorbent. This concentration gradient causes the substance to be transported from the environment into the zone with the lower concentration. However, this process is slower than transport of the substance to the surface of the sampler, so that in the immediate vicinity of the diffusion sampler the concentration of the substance is lower than in the environment. This fact leads to less substance being adsorbed than is predicted by Fick's first law of diffusion.

Investigation of the relationship between transport of the substance and the velocity of the air currents yielded the curves shown in Figure 2 [10].

This undesirable effect can be avoided if constant and sufficient transport of the substance to the surface of the sampler is guaranteed by an appropriate air current. The diffusion sampler must, however, be constructed in such a way that convection currents cannot enter the sampler, as this leads to uncontrolled movement of the substance [12]. There is a linear relationship between transport of the substance and the velocity of the air currents (Fig. 2, curve b). To achieve behaviour of the sampler which is as ideal as possible, the sampling phase is often protected by porous layers

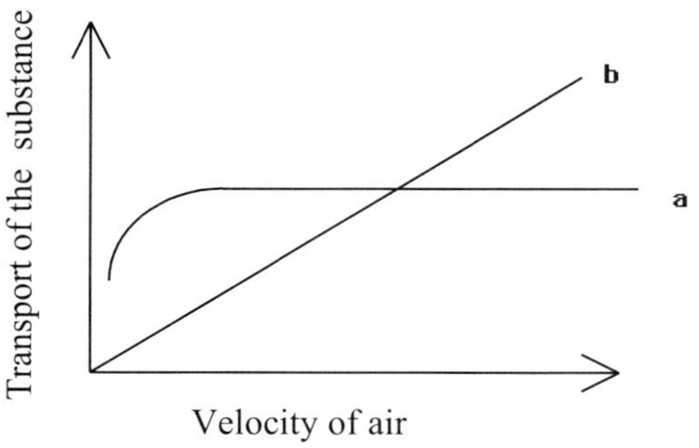

Fig. 2. Diagram of the effect of convection currents on transport of a substance for two types of samplers.
a = Curve for samplers protected against the entry of convection currents.
b = Curve for samplers which convection currents can enter.

or covers, for example by wire mesh, thin and porous synthetic membranes and thick permeable membranes. All of these measures serve to impede diffusion as little as possible while presenting convection currents with as large a resistance as possible, so that transport of the substance to the sampling phase occurs exclusively by diffusion and not by convection.

Another way of avoiding transport of the substance by convection lies in the geometry of the sampler, i. e. in the ratio of length to cross-section of the openings. Transport of the substance by diffusion is influenced on the one hand by the resistance to diffusion presented by the porous cover, and on the other also by the resistance to diffusion within the sampler and the sampling phase. Decisive is always the greater value. If the diffusion chamber is relatively short with a large cross-section, the resistance to diffusion inside the sampler is negligible, i. e. the resistance to diffusion of the cover is decisive for transport of the substance. If the diffusion chamber behind the cover is longer with a small cross-section, the resistance within the sampler is accordingly increased and thus becomes the decisive factor.

The least influence of convection currents and the least deviation of the experimentally determined concentrations from the theoretical value was observed by Pozoli and Cottica [10] when the two sources of resistance to diffusion were approximately equal. For the construction of a diffusion sampler, the anti-convective resistance of the porous cover must therefore be taken into consideration when calculating the length and the cross-section of the diffusion chamber.

When diffusion samplers are used in practice, a minimum velocity of air currents is required to ensure a constant adsorption rate, as shown in Figure 2, curve a. This is shown also in the results of Blome and Hennig [11]. For most of the types of samplers available commercially, the minimum velocity of the air currents is in the range from about 0.1 to 0.2 m/s. The position of the sampler relative to the direction of the air currents is not seen to have an effect at low wind speeds. The manufacturer's instructions on this point should be observed.

1.1.3.5 The effect of the duration of exposure and the concentration of the substance

The duration of exposure and the concentration of the substance have a great influence on the constancy of the rate of adsorption. The rate of adsorption can only remain constant for as long as the difference between the concentration of the substance in the environment and that on the surface of the adsorbent in the sampler is unchanged. When the concentration of the substance in the environment is constant, this concentration gradient is dependent only on the amount of substance already adsorbed and on the capacity of the sampling phase (see Section 1.1.2). If the sampling phase is exposed until its capacity is exhausted, the rate of adsorption decreases, or desorption and back diffusion of the collected substance into the workplace atmosphere takes place. Also very high concentrations, which initially cause a high rate of adsorption, lead to rapid exhaustion of the capacity of the sampling phase and then to a corresponding decrease in the rate of adsorption. Both factors – exposure duration and substance concentration – are therefore responsible for the constancy of the rate of adsorption and thus for a reliable result. It therefore makes sense to establish the relationship between the product of exposure time t and concentration of the substance c and the rate of adsorption. The optimum is a product $t \cdot c$ which is as high as possible at a constant rate of adsorption. This can be achieved by using a large amount of adsorbent with a sampler whose geometry ensures a low rate of adsorption. Such samplers have the disadvantage, however, that they react only very slowly to fluctuations in concentration (see Section 1.1.3.1).

1.1.3.6 The effect of substance-specific properties

In adsorption samplers, not only the substance-specific diffusion coefficient but also the behaviour of the substance towards the sampling phase is very important for the results of the analysis.
Substance-specific properties, such as e.g. molecule size, polarity and boiling point determine the adsorbed weight of substance per weight of adsorbent. If several substances are present simultaneously, competition can occur during adsorption. As a result of the substance-specific properties mentioned above, some substances are adsorbed to a greater degree than others, i.e. substances which are very strongly bound (high adsorption enthalpy), can displace others, so that desorption is observed.
The behaviour of the diffusion sampler towards mixtures must therefore be tested experimentally. In the case of reaction samplers (type A), the specificity of the reagent or the levels of interfering reactions is the limiting factor.

1.2 The construction of diffusion samplers

Passive samplers are available in various models, which, depending on the sampling to be carried out, can have advantages or even disadvantages. These specific properties must be taken into consideration when planning sampling and when using the various types of passive samplers. Adsorption samplers and reaction samplers must be distinguished. Adsorption samplers are merely systems for collecting samples, which must subsequently be analysed in the laboratory. Reaction samplers are available both as sampling systems and as analytical systems with which the results can be directly evaluated at the site of sampling, e.g. by a change in the colour of an indicator. The diagram below shows the types of passive sampler available.

1.2.1 Adsorption samplers

Adsorption samplers are made up of a sampling phase (e.g. activated carbon) mounted a short distance behind an opening with a known cross-section which is in contact with the ambient air. The molecules are transported from the opening to the collection phase by controlled diffusion.
At present the following types of adsorption sampler are available commercially:

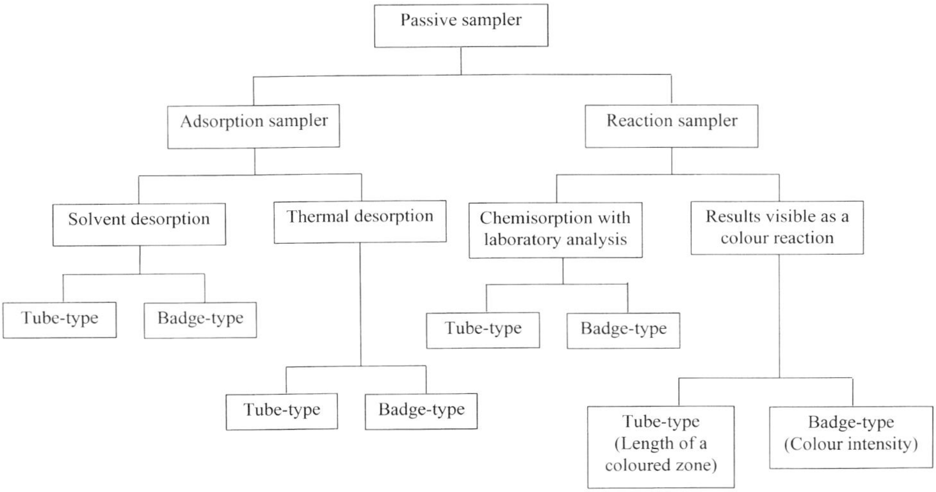

Fig. 3. Types of passive samplers classified according to DIN EN 838 [2].

- Tube-type (e.g. glass) with 2 openings over the cross-sections and diffusion barrier for sampling, enrichment of volatile organic compounds on activated carbon, liquid desorption, and analytical determination [13].
- Tube-type (e.g. glass, metal) with one opening over a cross-section for sampling, enrichment of volatile organic compounds on an organic polymer (e.g. Tenax), thermal desorption, and analytical determination [14].
- Tube-type (e.g. glass) with one opening over a cross-section for sampling, enrichment of nitrous oxide (laughing gas) on a molecular sieve, thermal desorption, and analytical determination [15].
- Badge-type (e.g. plastic) with a large inlet (membrane as diffusion barrier) for sampling, enrichment of volatile organic compounds on activated carbon, liquid desorption, and analytical determination [13].
- Badge-type (e.g. metal) with a large inlet (membrane as diffusion barrier) for sampling, enrichment of volatile organic compounds on an organic polymer (e.g. Tenax), thermal desorption, and analytical determination.

Whereas commercially available passive samplers for liquid desorption are ready to use and do not need any pretreatment, passive samplers for thermal desorption must usually be conditioned by heating in the laboratory before sampling.
The main features of the various types of adsorption sampler are summarised below.

1.2.1.1 Tube-type adsorption samplers (liquid desorption)

- Sampling rate lower than that of badge-type samplers; therefore lower sensitivity (generally this is not of importance for sampling at the workplace)
- A minimum wind speed of 1 to 2 cm/s is necessary

- Adsorption capacity is high relative to the sampling rate; therefore the constancy of the sampling rate is usually guaranteed even with high substance concentrations and long sampling periods
- The response time is 1 to 2 seconds (therefore even short changes in concentration are recognised)
- The sampler can only be used once
- Liquid desorption (analysis can be repeated several times)

1.2.1.2 Badge-type adsorption samplers (liquid desorption)

- Sampling rate higher than that of tube-type samplers; therefore greater sensitivity
- A minimum wind speed of 10 to 20 cm/s is necessary
- Adsorption capacity is low relative to the sampling rate; therefore the constancy of the sampling rate is not always guaranteed with high substance concentrations and long sampling periods
- The response time is 5 to 10 seconds (concentration peaks which last for a shorter period are not recognised)
- The sampler can only be used once
- Liquid desorption (analysis can be repeated several times)

1.2.1.3 Tube-type adsorption samplers (thermal desorption)

- Sampling rate lower than that of tube-type samplers with liquid desorption; high sensitivity as the whole sample is analysed at once (no dilution effect due to liquid desorption agents)
- A wind speed of <1 cm/s is sufficient
- Constancy of the sampling rate depends on the adsorbent
- Various adsorbents may be used, e.g. Tenax, Porapak, Chromosorb
- The adsorption capacity depends on the adsorbent. The constancy of the sampling rate must therefore be tested for various substance concentrations and sampling periods
- The response time is a few seconds (therefore even short changes in concentration are recognised)
- The sampler can be re-used after analysis
- Thermal desorption allows only one analysis per sample

1.2.1.4 Badge-type adsorption samplers (thermal desorption)

- Sampling rate higher than that of tube-type samplers; corresponds to that of badge-type samplers for liquid desorption
- High sensitivity as the whole sample is analysed at once (no dilution effect due to liquid desorption agents)

- A minimum wind speed of 10 to 20 cm/s is necessary
- Short sampling periods are possible. It must be remembered that the sampling rate is not linear; even with longer sampling periods and in particular with high substance concentrations the constancy of the sampling rate must be checked
- The response time is 5 to 10 seconds (concentration peaks which last for a shorter period are not recognised)
- The sampler can be re-used after analysis
- Thermal desorption allows only one analysis per sample

1.2.2 Reaction samplers

Unlike adsorption samplers, with which the substance to be determined is enriched on the sampling phase only as a result of physical adsorption and remains unchanged, reaction samplers enrich and immobilise the sample by chemical transformation of the substance. The enrichment phase containing the chemical reagents can be made of granulate (e.g. silica gel), strips of paper, wire mesh or solutions. With reaction samplers which display the results directly, the reaction product causes a change in the colour of an indicator, whose colour intensity (for badge-type samplers) or length of the coloured zone (for tube-type samplers) is evaluated after sampling. In addition to reaction samplers with a chemical colour reaction, samplers are available which work on an enzymatic basis. Reaction samplers which do not display the result directly must be analysed in the laboratory after sampling, like the adsorption samplers.
At present, the following types of reaction sampler are available commercially:

- Reaction samplers with an inlet for sampling, filled with absorption liquid; in general analysis in the laboratory is necessary.
- Tube-type samplers with an inlet over the cross-section for sampling, chemisorption on impregnated wire mesh; desorption; reaction; in general analysis in the laboratory is necessary [16].
- Badge-type samplers which display the results directly; evaluation of the colour intensity.
- Tube-type samplers which display the results directly, made of a glass tube with an opening for sampling, filled with granulated indicator material or impregnated strips of paper; evaluation of the length of the coloured zone.
- Reaction samplers on an enzymatic basis, evaluation of the colour intensity.

The main features of the various types of reaction sampler are summarised below.

1.2.2.1 Badge-type reaction samplers (laboratory analysis)

Transport of the substance from the ambient atmosphere to the enrichment phase (absorption solution) takes place e.g. via a membrane. After sampling, reagents, which may be integrated in the sampler, are added. The colour intensity is evaluated spectrometrically. The possibility of interfering components (cross-reaction) must be taken into consideration.

1.2.2.2 Tube-type reaction samplers (laboratory evaluation)

Transport of the substance from the ambient atmosphere to the enrichment phase, which is made of e. g. impregnated wire mesh, takes place via a diffusion path within a tube. After sampling, the wire meshes are eluted and reagent solutions are added. Evaluation is carried out spectrometrically. The possibility of interfering components (cross-reactions) must be taken into consideration. The tube and wire mesh can be used repeatedly. Reaction samplers of this type are used e. g. for monitoring the NO_2 concentration in the outdoor ambient air [17].

1.2.2.3 Badge-type reaction samplers (evaluation of the colour intensity)

Reaction samplers of this type yield a comparatively high sampling rate as they have a large diffusion cross-section and a short diffusion path to the reaction phase. This leads to great sensitivity. Chemical transformation to produce a coloured reaction product takes place in the reaction phase. Evaluation is carried out visually e. g. by comparison with a colour standard. With high substance concentrations the reagent is completely used up within a short period. Evaluation of the colour intensity is then no longer possible. The possibility of interfering components (cross-reactions) must be taken into consideration.

1.2.2.4 Tube-type reaction samplers (with a scale for the length of the coloured zone)

The sampling rate is a function of the coloured zone formed in the tube. To start with, the sampling rate is comparatively high as the diffusion path to the carrier material with unused reagent tends towards zero. As the length of the coloured zone increases, the diffusion path becomes longer, which results in a reduction in the sampling rate. Evaluation is carried out visually using the length of the coloured zone formed and a scale printed on the tube. The calibration curves take the form of a parabola [18]. The possibility of interfering components (cross-reactions) must be taken into consideration.

1.2.2.5 Reaction samplers on an enzymatic basis (evaluation of the colour intensity)

Reaction samplers on an enzymatic basis are similar to badge-type samplers with regard to the sampling rate and evaluation. Instead of a chemical reaction, such systems are based on biochemical reactions which lead to a colour change in the reaction phase. Reaction samplers on an enzymatic basis are characterized by high sensitivity and selectivity. Reaction samplers of this type are used e. g. for estimating formaldehyde concentrations [19].

1.3 Demands made of diffusion samplers

Workplace air is monitored to check that threshold limit values are observed. Both the mean values for the shift and short-term exposure peaks must be determined. The procedure for carrying out workplace monitoring described in TRGS 402 [20] makes certain demands on the monitoring procedure which must also be fulfilled when using diffusion samplers.
To make it easier for the user to decide about the use of a diffusion sampler, the requirements and test procedures for diffusion samplers under prescribed laboratory conditions are laid down in the standard DIN EN 838 [2]. It is the responsibility of the manufacturer of the diffusion sampler to carry out this suitability test. The performance characteristics tested must include the inaccuracy associated with the analysis as stipulated in DIN EN 482 [21].

1.3.1 Standard requirements according to DIN EN 838

The requirements laid down in DIN EN 838 [2] for diffusion samplers apply both for samplers used for direct determination of the concentration (type A, e. g. diffusion tube with colour zone display), and also for all types of sampler used for indirect determination of the concentration with sampling and analysis in separate steps (type B). The latter are sub-divided into types B1 (adsorption on a solid phase and desorption using a solvent with subsequent analysis of the desorbed substance), B2 (adsorption on a solid phase and desorption by heating, with subsequent analysis of the desorbed substance) and B3 (absorption in a liquid with subsequent analysis of the solution). Standard requirements are summarized in Table 1.

Table 1. Standard requirements.

Parameters	Requirements	Comments
Desorption efficiency	Type B1: ≥ 0.75 with a coefficient of variation of ≤ 0.1 Type B2: ≥ 0.95 with a coefficient of variation of ≤ 0.1	applies for each use of the sampler
Shelf-life of the loaded samplers	Type B: the mean recovery after storage should not differ from that before storage by more than 10%	shelf-life of 2 weeks at room temperature or according to the manufacturer's instructions
Rate of adsorption	Type B: if it can be calculated according to equation (5), the nominal value must be within ± 25 % of the theoretical value	
Labelling the sample	a suitable surface must be available for the user to label the sample	

Table 1. (continued)

Parameters	Requirements	Comments
Temperature range	5 °C to 40 °C, but at least from 10 °C to 30 °C	the requirements for analytical error must be fulfilled in this range (correction factors only permissible outside the range 10 °C to 30 °C)
Relative humidity	20% to 80%	the requirements for analytical error must be fulfilled in this range (without the use of correction factors)
Blank value	less than a third of the calculated rate at which the sampler adsorbs the substance during an exposure period of 30 minutes and at an exposure concentration of 0.1 of the threshold limit value	determined using 6 unspiked samplers
Sampler integrity	the additional concentration of the substance to be analysed determined above the blank value must be less than a third of the calculated rate at which the sampler adsorbs the substance during an exposure period of 30 minutes and at an exposure concentration of 0.1 of the threshold limit value	determined using closed samplers of type B, which were exposed for 4 hours to an atmosphere of the substance to be determined at twice the threshold limit value
Wind speed and position of the sampler	wind speed varies between 0.01 m/s and 4.0 m/s; positioning parallel to or at right angles to the air current	samplers only for personal air sampling worn by the individual: must be tested at 0.1 m/s to 1.5 m/s for indoor workplaces and at 0.1 m/s to 4.0 m/s for indoor workplaces and outdoor workplaces
Shelf-life (closed unloaded samplers)	Type A: at the end of the storage period the results should not differ from the original results by more than 10%	the shelf-life of the samplers in the original packaging must correspond with the manufacturer's declaration
Mechanical durability	test with prescribed test apparatus	
Inaccuracy associated with the analysis	must lie within the requirements of DIN EN 482 for all types	

The diffusion samplers corresponding to this standard are grouped in classes 1 A and 1 B. Samplers in class 1 A have been tested according to the standardising part of this standard or in some cases by the standardising multi-factor plans contained in this standard. Samplers in class 1 B, which only applies for samplers of type B, are tested using a substance from a homologous series, for which previously both lower and higher members have been tested and found to correspond to class 1 A. For class 1 B at least the desorption efficiency and the diffusive adsorption rate must be determined.

1.3.2 Demands made of diffusion samplers regarding the inaccuracy associated with the analysis

In the European standard DIN EN 482 [21] general requirements for analytical procedures for workplace monitoring have been formulated, and must also be fulfilled by diffusion samplers. It must be remembered that the requirements apply to the whole analytical procedure, i.e. for sampling with the diffusion sampler, and the usual preparation of the sampling phase and analytical determination.

Table 2. Requirements of analytical procedures which use diffusion samplers, defined according to the task [21].

Task	Relative inaccuracy associated with the analysis	Minimum range	Duration of averaging
Determination of the mean concentration	≤ 50%	0.1 to 0.5 times the threshold limit value	≤ reference time for the threshold limit value
Sampling near the source of an emission	≤ 50%	0.5 to 10 times the threshold limit value	depends on the source
Comparison with the threshold limit value	≤ 50%	0.1 to 0.5 times the threshold limit value	≤ reference time for the threshold limit value
	≤ 30%	0.5 to 2 times the threshold limit value	
Control samples	≤ 50%	0.1 to 0.5 times the threshold limit value	≤ reference time for the threshold limit value
	≤ 30%	0.5 to 2 times the threshold limit value	

The other requirement of DIN EN 482, that of obtaining an overview of the temporal and/or spatial exposure situation via the concentration distribution, can often not be met with diffusion samplers for which sampling periods of a maximum of 5 or 15 minutes for averaging are prescribed. In such cases, instruments with a direct display (e.g. flame ionisation detector, electro-chemical sensors) or sampling procedures with active sampling should be used.

The relative inaccuracy associated with the analysis can be calculated as follows:

$$\frac{|\bar{x} - x_{\text{ref}}| + 2\,s}{x_{\text{ref}}} \cdot 100 \qquad (8)$$

where:

$\bar{x}$ is the mean of the results in mg/m^3 from n repeated analyses

x_{ref} is the true value or the assumed reference value of the concentration in mg/m^3

s is the standard deviation of the analytical results in mg/m^3

The relative inaccuracy associated with the analysis, must, according to the task, be determined at the lower and upper end of the minimum range shown in Table 2 and for at least one intermediate concentration. It is recommended that the third concentration chosen is at the level of the threshold limit value in air. The number of repeated analyses at each of these concentrations must be at least six.

Testing the analytical procedure can, at present, only be carried out under laboratory conditions, as the generation of a test gas atmosphere of known concentration is the prerequisite for the determination of the relative inaccuracy associated with the analysis [5]. The minimum range prescribed can generally be covered by diffusion samplers by varying the sampling period (averaging period). Limitations are most likely at the lower end of the minimum range if it cannot be guaranteed that the minimum amount can be collected within the period set for averaging concentrations for the threshold limit value.

If a diffusion sampler does not fulfil the requirements for the relative inaccuracy associated with the analysis and for the minimum range, it cannot be used for the sampling task in question.

The requirements of this standard for the specificity and selectivity of diffusion samplers can only be applied to samplers of type A. They must yield a clear result for the concentration in the given range. Interfering components must be known. If the colour zone display of the diffusion tube also reacts to substances in the workplace atmosphere other than those to be determined, this sampler may only be used if the signal is increased by the interfering components. The whole value must then be assigned to the substance to be determined. If as a result of interfering components the values obtained are too low, the sampler is not suitable for this task.

Samplers of type B, with which sampling is usually followed by an analytical procedure in the laboratory, are allowed also to detect substances other than those to be determined. The analytical step must, however, allow the separation of these substances and the clear assignment of the signals. These samplers can be used with the restriction that interfering components must not compete with the substance to be determined. This means that the substance to be determined may not be displaced from the sampling phase by interfering components. In addition, the total amount of all the substances in the workplace air must not exceed the capacity of the sampler.

1.3.3 Determination according to TRGS 402

Diffusion samplers can be used for analyses carried out in work areas to determine whether the threshold limit values are being observed and also for control analyses; observance of the shift average value and the peak value must be monitored.
Diffusion samplers which conform to DIN EN 838 [2] are in principle suitable for monitoring the shift average values. An averaging duration of an hour or more should be used. If the averaging duration is shorter, which means an increased number of samples, the sampling rate of the sampler must be checked beforehand.
Diffusion samplers can only be used with reservations for monitoring short-term exposure peaks. Usually 15-minute average values are determined in this case. If sampling is carried out for less than one hour it must be checked whether the selected diffusion sampler meets the requirements of the analytical procedure with regard to the quantification limit and the concentration range of the substance to be monitored. It must then be decided whether the sampler can be used or not by comparing the data with the threshold limit value to be monitored. The relative inaccuracy associated with the analysis given in DIN EN 482 [21] must be taken into account.
In addition, with shorter sampling times the use of diffusion samplers is restricted by the fact that a procedure for workplace analyses must allow concentrations to be determined at least in the range from one tenth (or at least one fifth) of the threshold limit value to three times of the threshold limit value.

1.4 Analytical use of diffusion samplers

1.4.1 Sample collection

Diffusion samplers are sampling systems, which begin to sample substances as soon as the caps are removed. Before sampling, the sampling strategy is laid down according to TRGS 402. The diffusion sampler is opened at the beginning of sampling. The parameters which are important for the determination of the concentrations in air (date, time, temperature, atmospheric pressure and relative humidity) are noted in a sampling protocol. To determine the shift average value the sampling duration must be recorded to the nearest minute. To monitor the peak values the sampling duration must be recorded to the nearest second.
Sampling is carried out in the breathing zone for personal sampling. When selecting static positions for the sampler, the minimum wind speed given by the manufacturer must be taken into account. The opening of the diffusion sampler must not be obstructed. After sampling, the diffusion sampler is closed with suitable caps. The samples should be analysed immediately. If the samples are stored for a longer period of time until evaluation, they must be sealed with suitable caps. The transport and storage conditions depend on the type of diffusion sampler and the substances to be analysed.
When diffusion samplers are used, even unskilled persons can equip the test persons with the samplers according to the instructions of the analytical laboratory, close the

samplers after sampling and dispatch them according to the transport conditions stipulated by the laboratory.
Calibration, as necessary for personal sampling pumps, is not needed with diffusion samplers.
Systems are used in which the analysis is carried out in external laboratories or a colour reaction is recorded.

1.4.2 Preparation of the samplers

The samplers are prepared in accordance with the analytical procedure to be used. The most usual processes are thermal desorption and the use of liquid desorption agents.

1.4.2.1 Thermal desorption

The preparation of thermal desorption tubes takes place in the thermal desorber, without an extraction step. The diffusion samplers are put into an appropriate thermal desorber and heated; the substances collected are transferred with a carrier gas to a packed cold trap. When desorption is complete, the cold trap is heated very quickly so that the substance reaches the GC column as a concentrated band.
The conditions for the thermal desorber ATD-400 from Perkin Elmer which are recommended for the determination of solvent vapour are given here as an example (Tab. 3).

Table 3. Recommended parameters for thermal desorption from various sampling phases.

Adsorbent	Tenax TA	XAD 4, Chromosorb 106
Desorption temperature	250 °C	170 °C
Desorption time	10 minutes	5 minutes
Temperature of the transfer line	100 °C	100 °C
Cold trap (adsorption)	–30 °C	–30 °C
Cold trap (injection)	300 °C	300 °C
Weight of the adsorbent in the cold trap	20 mg Tenax TA	20 mg Tenax TA
Carrier gas	Helium	Helium
Input split		40 mL/min (input split)
Desorb flow		10 mL/min (desorb flow)
Output split		30 mL/min (output split)

The instrumental conditions have to be chosen according to the configuration of the apparatus if other types of thermal desorbers are used. After setting up the thermal desorber and the gas chromatograph, the calibration standards and the samples are analysed. Thermal desorption in gas cuvettes is also possible. The analytical determination is carried out by FTIR spectrometry [15].

1.4.2.2 Solvent desorption

Solvent desorption is used mainly with diffusion samplers filled with activated carbon. The complete contents of the diffusion sampler are transferred to a glass vessel and covered with a desorption agent. An internal standard is usually added. The adsorbed substances are desorbed by the solvent. The desorption efficiency must be determined using calibration standards. The humidity during sampling is a very important factor. Two procedures are used for the analytical determination.

a) Injection of the solvent

After desorption, the solvent must be separated from the adsorbent layer and transferred to glass bottles. Back diffusion onto the adsorbent is thus prevented. A µl-aliquot is then injected into an analytical system such as e. g. a gas chromatograph or a liquid chromatograph. Carbon disulfide, diethyl ether or mixtures of two or three solvents have proved to be reliable.

b) Headspace analysis

The whole sample – made up of adsorbent and desorption agent – is brought to the right temperature and the headspace above the sample is injected into a gas chromatograph. It is not necessary to separate the adsorbent and desorption agent. The boiling point of the solvent must be sufficiently high. For this procedure, desorption agents such as benzyl alcohol, dimethylacetamide, dimethylformamide and phthalic acid dimethylester are mainly used.

1.4.2.3 Reaction samplers

With reaction samplers, a chemical reaction takes place between the substance to be determined and a chemical in the sampler.
In practice, the reaction samplers used are either Draeger-tubes constructed as diffusion samplers from which the results are read immediately after sampling, or samplers with reagent solutions which must be analysed later in the laboratory. The adsorption agents are either reagent solutions or reactive sampling phases. The samples are usually evaluated photometrically.

1.4.3 Use of the diffusion samplers

Diffusion samplers are selected according to practical considerations.
If the substances are present as mixtures of vapours, chromatographic or spectrometric analytical procedures are used. These procedures always require that enrichment and sample preparation be carried out. Numerous substances can be determined with either solvent desorption or thermal desorption.
Diffusion sampling has proved reliable in these cases.

Whenever the substances are also present in particulate form, an apparatus suitable for sampling aerosols is required (e.g. SILPP for spray painting [22]). A diffusion sampler is placed behind a glass fibre filter in the sampling current. If the particulate phase cannot be separated from the vapour phase, diffusion samplers cannot be used.
Reaction samplers have proved reliable for inorganic gases and vapours, and for formaldehyde and some halogenated hydrocarbons.

1.5 Summary of areas of use

1.5.1 Analytical procedures for diffusion samplers

Passive sampling is gaining in importance relative to active enrichment procedures. A few tested methods of quantitative evaluation are listed in Tables 4 and 5.

Table 4. DFG methods [23].

Substances to be determined	Source	Sampling phase	Analytical procedure
Halogenated narcosis gases	DFG, No. 1	XAD-4	Thermal desorption/GC
Halogenated narcosis gases	DFG, No. 2	Activated carbon	Liquid desorption/GC
Styrene	DFG, No. 3	Tenax TA	Thermal desorption/GC
Nitrous oxide	DFG, No. 3	Molecular sieve 5Å	Thermal desorption/GC
Nitrous oxide	DFG, No. 2	Molecular sieve 5Å	Thermal desorption/FTIR

Table 5. Other methods of quantitative determination listed in the literature [13].

Substances to be determined	Source	Sampling phase	Analytical procedure
Acrylonitrile	HSE, MDHS 55	Porapak N	Thermal desorption/GC
Benzene	HSE, MDHS 50	Porapak Q	Thermal desorption/GC
1,3-Butadiene	HSE, MDHS 63	Molecular sieve 13X	Thermal desorption/GC
Volatile organic compounds (solvent mixtures)	HSE, MDHS 88	Activated carbon	Liquid desorption/GC
Volatile organic compounds (solvent mixtures)	HSE, MDHS 80	Tenax TA, Chromosorb 106	Thermal desorption/GC
Formaldehyde	HSE, MDHS 78	Silica gel (coated)	Liquid desorption/HPLC
n-Hexane	HSE, MDHS 74	Activated carbon	Liquid desorption/GC
Hydrocarbons	HSE, MDHS 66	Tenax TA, Chromosorb 106, Spherocarb	Thermal desorption/GC
Styrene	HSE, MDHS 43	Tenax TA	Thermal desorption/GC
Styrene	HSE, MDHS 44	Activated carbon	Liquid desorption/GC
Toluene	HSE, MDHS 64	Activated carbon	Liquid desorption/GC
Toluene	HSE, MDHS 69	Activated carbon	Liquid desorption/GC

1.5.2 Suitability of special sampling phases

Various adsorbents can be used for sampling organic substances. The methods described in Tables 4 and 5 are based mainly on these adsorbents.
Table 6 lists the areas of use of various sampling phases for the passive sampling of various substances and groups of substances. The selectivity of some adsorbents is described in [24].

Table 6. Adsorbents for passive sampling.

Adsorbent	Substances
Activated carbon	Solvent *
Carbopack B	Aliphatic and aromatic hydrocarbons
Carbotrap	Perfluorodimethylcyclobutane
Chromosorb 102	Freons, halogenated hydrocarbons
Chromosorb 106 or XAD-4	Aliphatic and aromatic hydrocarbons, halogenated hydrocarbons, esters, glycol ethers, alcohols, nitriles, ketones, glycidyl ethers
Graphitised carbon molecular sieves	Halogenated hydrocarbons, alcohols, carbon disulphide, dioxane, ethylene oxide
Porapak N	Nitriles
Porapak Q	Aromatic hydrocarbons, acrylates, glycol ethers
Molecular sieve 13X	1,3-Butadiene
Molecular sieve 5Å	Laughing gas
Tenax GR	Aromatic hydrocarbons, halogenated hydrocarbons, glycol ethers, ketones
Tenax TA	Aliphatic and aromatic hydrocarbons, halogenated hydrocarbons, esters, glycol ethers, aldehydes, alcohols, nitriles, glycidyl ethers, terpenes

* 2-Butanone cannot be stored for long after adsorption on activated carbon.

1.5.3 Reaction samplers

Reaction samplers can be used for the direct determination of the substance concentration. Table 7 shows a selection of currently available reaction samplers.

Table 7. A selection of reaction samplers.

Reaction samplers for the determination of	Manufacturer/ supplier*	Reaction samplers for the determination of	Manufacturer/ supplier*
Acetaldehyde	4	Carbon dioxide	2, 4
Acetone	4	Carbon monoxide	2, 4
Formic acid	4	Methyl isobutyl ketone	4
Ammonia	2, 3, 4	Olefins	2
Hydrogen cyanide	2, 4	Perchloroethylene	2, 4
1,3-Butadiene	2, 4	Phosgene	2
2-Butanone	4	Phosphine	2
Chlorine	4	Nitric acid	4
1,2-Dichloroethylene (*trans*)	4	Hydrochloric acid	2, 4
Dimethylamine	4	Sulfur dioxide	2, 4
N,N-Dimethylethylamine	4	Hydrogen sulfide	1, 2, 3, 4
Acetic acid	2, 4	Nitrogen dioxide	2, 3, 4
Acetic anhydride	4	Toluene	2, 4
Ethanol	2, 4	Trichloroethylene	2, 4
Ethene	4	Triethylamine	4
Ethyl acetate	2	Vinyl chloride	4
Formaldehyde	2, 3, 4	Vinylidene chloride	4
Hydrogen fluoride	4	Water vapour	2
Furfural	4	Hydrogen peroxide	4
Hydrazine	4	Xylene	4
Isoprene	4		

* Manufacturer/supplier: 1 Auergesellschaft, Berlin
2 Dräger Sicherheitstechnik GmbH, Lübeck
3 Günter Karl OHG, Gau-Algesheim
4 MTC Messtechnik-Chemie GmbH, Müllheim

1.6 References

[1] *Blome H, Hennig M* (1985) Leistungsdaten ausgewählter Passivsammler, 1. Part. Staub-Reinh. Luft 45: 505–508

[2] *Europäisches Komitee für Normung* (CEN) (1995) DIN EN 838-Luftbeschaffenheit am Arbeitsplatz-Diffusionssammler für die Bestimmung von Gasen und Dämpfen-Anforderungen, Prüfung, Kennzeichnung. Brüssel. Beuth-Verlag, Berlin

[3] *Bartholomé E, Biekert E, Hellmann H, Ley H, Weigert W, Weise W* (Eds) (1972) Ullmanns Encyklopädie der technischen Chemie. Vol 1, 4th ed., VCH-Verlag, Weinheim, p. 134

[4] *Näser KH* (1960) Physikalische Chemie für Techniker und Ingenieure. 16. ed., VEB Deutscher Verlag für Grundstoffindustrie, Leipzig, p. 359

[5] *Nelson GO* (1971) Controlled Test Atmospheres-Principles and Techniques. Ann Arbor Sci. Publ., Inc., Ann Arbor, Michigan

[6] *Einfeld W* (1983) Diffusional sampler performance under transient exposure conditions. Am. Ind. Hyg. Assoc. J. 44: 29

[7] *Lautenberger WJ, Kring EV, Morello JA* (1980) A new personal badge monitor for organic vapors. Am. Ind. Hyg. Assoc. J. 41: 737–747

[8] *Landesumweltamt Nordrhein-Westfalen* (Eds) (1998) Validierung von Passivsammlern für Immissionsmessungen von Kohlenwasserstoffen. Materialien Nr. 46, Essen

[9] *Ikeda M, Koizumi A, Kashora A* (1987) Validation of passive dosimetry through biological monitoring and its application in solvent workplaces. In: Berlin A, Brown RH, Saunders KJ (Eds) Diffusive sampling-An alternative approach to workplace air monitoring. The Proceedings of an International Symposium, Luxembourg, 22nd-26th Sept. 1986. Publication No. 10555 EN, Royal Society of Chemistry, London, p. 14

[10] *Pozoli L, Cottica D* (1987) An overview of the effects of temperature, pressure, humidity, storage and face velocity. In: Berlin A, Brown RH, Saunders KJ (Eds) Diffusive sampling-An alternative approach to workplace air monitoring. The Proceedings of an International Symposium, Luxembourg, 22nd-26th Sept. 1986. Publication No. 10555 EN, Royal Society of Chemistry, London, p. 119

[11] *Blome H, Hennig M* (1985) Leistungsdaten ausgewählter Passivsammler. Staub-Reinh. Luft 45: 541–546

[12] *Pannwitz KH* (1981) Orsa 5-Ein neuer Probenehmer für organische Lösemitteldämpfe. Drägerheft 321: 1–6

[13] *Health and Safety Executive* (HSE) (Ed) (1999) Occupational Medicine and Hygiene Laboratories: Methods for the Determination of Hazardous Substances. MDHS No. 1–95, HSE Books, Sudbury, Suffolk

[14] *Greim H* (Ed.) (1998) Analytische Methoden zur Prüfung gesundheitsschädlicher Arbeitsstoffe, Luftanalysen. Vol 1, Lösemittelgemische Meth.-Nr. 5. WILEY-VCH Verlag, Weinheim

[15] *Greim H* (Ed.) (1998) Analytische Methoden zur Prüfung gesundheitsschädlicher Arbeitsstoffe, Luftanalysen. Band 1, Distickstoffmonoxid Meth.-Nr. 2. WILEY-VCH Verlag, Weinheim

[16] *Palmes ED, Tomczyk C* (1979) Personal sampler for NO_x. Am. Ind. Hyg. Assoc. J. 40: 588–591

[17] *Hangartner M* (1988) Passivsammler für die Immissionsmessung von Stickstoffdioxid. Drägerheft 368: 6–8

[18] *Pannwitz KH* (1984) Anzeigende Diffusionsröhrchen. Drägerheft 328: 8–13

[19] *Schwarzer S, Plümke K* (1994) Bio-Check F-eignungsgeprüft mit TÜV-CERT-Zertifikat. Drägerheft 358: 9–11

[20] *Bundesministerium für Arbeit und Sozialordnung* (1997) TRGS 402 Ermittlung und Beurteilung der Konzentrationen gefährlicher Stoffe in der Luft in Arbeitsbereichen. In: Technische Regeln und Richtlinien des BMA zur Verordnung über gefährliche Stoffe. BArbBl. 11: 27–33

[21] *Europäisches Komitee für Normung* (CEN) (1994) DIN EN 482- Arbeitsplatzatmosphäre-Allgemeine Anforderungen an Verfahren zur Messung von chemischen Arbeitstoffen. Brüssel. Beuth Verlag, Berlin

[22] *Lichtenstein N, Hennig M, Friedrich C, Auffarth J, Hebisch R, Rentel KH, Fricke HH, Möcklinghoff K, Dahmann D* (1997) Meßmethoden zur Bestimmung der Exposition gegenüber Lackaerosolen und Lösemitteldämpfen beim Spritzlackieren. Staub-Reinh. Luft 57: 39–45

[23] *Greim H* (Ed.) (1998) Analytische Methoden zur Prüfung gesundheitsschädlicher Arbeitsstoffe, Luftanalysen. Band 1, 1.-11. Lfg., WILEY-VCH-Verlag, Weinheim

[24] *Stanetzek I, Giese U, Schuster RH, Wünsch G* (1996) Chromatographic characterization of adsorbens for selective sampling of organic air pollutants. Am. Ind. Hyg. Assoc. J. 57: 128–133

Authors: *U. Giese, R. Hebisch, K.-H. Pannwitz, M. Tschickardt*

Quality control

Contents

1 Quality control

The purpose of quality control is to check that the analytical procedure functions correctly during routine use. It is designed to guarantee that no unrecognised changes during analysis influence the analytical result. Specific checking and monitoring procedures must be prescribed so that the person responsible for the analysis can be sure that the whole analytical system always yields tolerable results.
With appropriate methods, important analytical, functional and statistical parameters are monitored to check that they remain within a given statistically determined range.

The monitoring should cover sample preparation, and, if possible, sampling.
Because the devices used today in instrumental analysis are equipped with data recording and processing systems, the quality control procedure must also monitor the hardware and software.

1.1 Analytical parameters

The parameters of importance for validating analytical results are described briefly below [1, 2]. Functional and statistical parameters must be distinguished. The functional parameters include in particular accuracy, sensitivity, recovery, selectivity and specificity. The statistical parameters include precision, the detection limit or quantification limit, as well as reproducibility and comparability, parameters associated with precision and standard deviation.

1.1.1 Accuracy

The accuracy of analytical results is influenced by systematic errors. They are determined by analysing reference samples. The certified values of the reference samples are, when possible, determined with different, independent analytical procedures.

1.1.2 Sensitivity

Another important parameter that can influence the validity of analytical results is the sensitivity of the analytical procedure. This is generally understood to be the differential quotient of the characteristic function of the procedure. For analytical procedures, the characteristic function corresponds to the analytical calibration function $x = f(c)$. Accordingly, the sensitivity is defined as dx/dc (slope of the calibration function). An analytical procedure is sensitive if a small change in concentration c results in a large change in the determined value x. The sensitivity does not need to be constant; only if the analytical calibration function is linear in the whole range is the sensitivity constant. It must, however, be reproducible and, being an important parameter, must always be monitored.

1.1.3 Selectivity and specificity

Two other concepts associated with the reliability of analytical procedures are selectivity and specificity. An analytical procedure is regarded as being completely selective if the various components (elements) in the analytical sample can be determined independently of one another with this procedure. In this strict sense there are no selective analytical procedures. Other components of the sample always influence the analysis (interference). Important is, however, that this influence is as small as possible, and in some cases not even detectable. Analytical procedures with which the interference is relatively small, are described as moderately selective. An analytical procedure is regarded as being specific if it responds to only one component, even if other components are present in the analytical sample. These parameters must also be checked regularly, if e.g. the sample has a different matrix, or other components are present.

1.1.4 Recovery

The recovery is a criterion for evaluating an analytical procedure or an individual step in the procedure. If a recovery of 100% is found when the individual steps of a procedure are checked, the procedure is free of systematic errors, both constant and proportional errors. If, however, proportional systematic errors were detected, future analyses must be carried out with the standard addition procedure. If constant or proportional systematic errors have been detected, a warning must be noted in the analytical method.

1.1.5 Statistical parameters

The group of statistical parameters is associated with the concept of random error. Random errors are caused by incalculable and unalterable changes during analysis in the apparatus, the sample to be analysed, the environment and the observer. If the same observer repeats an analysis with the same sample, determining the same parameter with the same apparatus under the same conditions, or if an observer compares several times the same apparatus with the same sample under the same conditions, the individual analytical values differ from each other, they are scattered over a range.
The random errors differ in their magnitude and sign, cannot be detected individually and make the result uncertain. They can, however, be estimated and characterised in total using suitable calculations; the greater the number of analyses carried out, the greater is the reliability of this estimation. The random errors may be described quantitatively in terms of the standard deviation, provided that the distribution of the results is a normal Gaussian distribution.

1.1.6 Reproducibility and comparability

Also associated with the random error of analytical procedures are the two parameters reproducibility and comparability. The definitions from ISO 5725 are given below [3].
Reproducibility is the value r below which one can expect with a prescribed probability (usually 95%) to find the absolute difference between two individual test results (analytical values) which were obtained with the same procedure with identical test material and the same conditions (employee, apparatus, laboratory, short time period). It is also referred to as the within-laboratory variance.

$$r = 2.83 \cdot t\ (S; n-1) \cdot s_r \qquad (1)$$

The factor t, for a prescribed certainty S and n determined values, is described in several literatures. Comparability is the value R below which one can expect the absolute difference between two individual test results (analytical values), which were obtained with the same test material, but under different conditions (employee, apparatus, laboratory and/or at different times), to be with a prescribed probability (usually 95%). It is also referred to as the between-laboratory variance.

$$R = 2.83 \cdot t\ (S; n-1) \cdot s_R \tag{2}$$

Other statistical parameters include the detection limit and quantification limit. As for the precision, for the detection limit of a complete analytical procedure, a numerical value which can be objectively determined and checked can be given. This value indicates which smallest concentration or amount can be reliably detected with this procedure (not, however, determined exactly). There is, of course, some risk involved in analysis of amounts close to the detection limit, but this risk can be estimated.

1.1.7 Detection limit

With low concentrations it is uncertain whether an observed value results from the element sought (in general, a substance), or whether it is caused by interference. It is not a question of the smallest concentration or amount still reliably detectable, but of which determined values may be recognised as "analytical signals" and which values must be discarded. To be able to answer this question, the whole analytical procedure must be carried out without the substance present and the average blank value x_{bl} and the standard deviation s_{bl} of the blank value be determined. It is important that the usual analytical procedure is used. This, of course, includes all the steps of sample preparation, the use of the same chemicals, the same vessels etc. The only difference should be that the analytical element is not detectable in the blank sample, which is of the same kind as the real sample. This can sometimes be a difficult problem. When all these conditions have been fulfilled, the mean value x_{bl} is determined; the number n of blank value analyses should not be less than 10. It is important to note that it is not the magnitude of the blank value itself which makes the evaluation of a determined value uncertain, but the size of its random fluctuations. Generally, for the detection limit

$$\underline{x} = \bar{x}_{bl} + k \cdot \sigma \tag{3}$$

To be able to calculate the detection limit $\underline{c}$ (concentration) or $\underline{q}$ (amount), the form of the analytical calibration function $x = g\ (c)$ or $x = g\ (q)$ for small amounts of the substance must be known. With $k = 3$ and from the sensitivity $\Delta A/\Delta c$ of the procedure, the detection limit $\underline{c}$ is given by

$$\underline{c} = 3 \cdot \left(\frac{\Delta A}{\Delta c}\right)^{-1} \cdot \sigma_{bl} \tag{4}$$

Analyses of amounts close to the detection limit are not very exact. If the value x or A is a linear function of the concentration c, with $k = 3$ the relative standard deviation of determinations at the detection limit is at least 33%.

1.1.8 Quantification limit

For concentrations close to the detection limit, the fluctuations in the values determined are very large. It can therefore occur that for a concentration

$$c = \underline{c} + \sigma \tag{5}$$

as a result of the random fluctuation, a value $x \leq \underline{x}$ is registered. This may not be considered to be real. In fact, 50% of the values which belong to concentration c are below $\underline{x}$. Thus it is clear that it would be false to conclude that when the determined value $x \leq \underline{x}$ the concentration in the sample is $c \leq \underline{c}$. Therefore a quantification limit c_{QL} is defined. With a given certainty, concentrations which are greater or equal to c_{QL}, provide signals which are larger than

$$\underline{x} = \bar{x}_{bl} + k \cdot \sigma \tag{6}$$

To guarantee that the quantification limit fulfils this requirement, the difference between the measured value and the mean blank value must be twice that for the detection limit. This means

$$c_{QL} = 6 \cdot \left(\frac{\Delta A}{\Delta c}\right)^{-1} \cdot s_{bl} \tag{7}$$

For practical reasons, instead of the factor 6 between the detection limit and quantification limit often the factor 10 is chosen.
Knowledge of the functional and statistical parameters discussed above, all of which have an influence on the result and thus on the validation of the analytical procedure or analytical results, is imperative. In daily routine practice, in particular

- mean values
- blank values (detection and quantification limits)
- recoveries
- ranges
- (sensitivities)

must be recorded and evaluated using suitable analytical procedures and documentation. It can be useful to follow up other parameters.

1.2 Quality control

Only validated analytical procedures are used.
Whenever possible, for analyses in air, validated procedures from the available collections of methods should be used [4].
If these procedures are adapted to suit the task in question, they must be checked by analysing internal control samples to establish whether they fulfil the requirements of EN 482 [2].

1.2.1 Testing analytical procedures

Validation must always be carried out for newly developed or modified procedures. This takes place either by

- comparison of the results of the analyses with those of a validated or
- an independent analytical procedure, or by
- analysis of reference materials or of
- control samples prepared in the laboratory itself.

In the above list the methods are listed in order of priority, the preferred methods first. The equal validity of the procedures is shown by testing for systematic differences between the analytical results.
In analyses of reference samples and prescribed control samples, the result of the analysis is compared with the given reference value.
The testing of an analytical procedure over a particular concentration range is carried out by analysing two identical series of samples with the new procedure and the independent or reference procedure. The results are then checked for systematic differences. If systematic differences are found, the results are investigated further and the reasons for these differences – if detected – remedied. The corrected procedure is then tested again for systematic differences. When no more systematic differences can be detected, the characteristic data of the method are determined. Then a detailed procedure which includes the results of this test is drawn up.
Analytical procedures for which the results showed systematic differences from those of a validated procedure are not used.
There are numerous aids for determining the control parameters; they are described briefly below.

1.3 Quality control cards and control samples

1.3.1 Control samples

For the validation of analytical results, reference materials or certified materials must be available. Certified reference materials, i.e. materials containing a confirmed concentration of analytes, are produced and distributed by internationally recognised organisations or institutions [5–7]. In analytical laboratories reference materials are used for calibration ("calibration standards") and as controls ("control standards"). Certified data alone, however, do not guarantee success; the reference materials must be used correctly. Depending on the samples to be analysed and the technique used, the correct understanding of the problem and appropriate choice of reference material is important.

Requirements for control material:

- representative with regard to the matrix and concentration
- the substance levels in the control material cover the analytically important ranges
- available in sufficient amounts
- stability over several months has been demonstrated
- not influenced by the storage vessel
- the removal of samples does not lead to changes in the remaining control material

Types of control material:

- *Standard solutions*
  (solution of a standard substance in a suitable solvent, as similar as possible to the sample)
- *Blank solutions*
  (sample to be analysed, free of the components to be determined, which is put through the whole analytical procedure)
- *Real samples*
  (in addition to the substance to be determined contains many unknown components, suitable for precision analysis)
- *Real doped samples*
  (with a defined added amount of the standard substance, for recovery controls)
- *Synthetic solutions*
  (in addition to the substance to be determined contain interfering components in known concentrations)
- *Certified standard solutions*
  (real or synthetic solutions with certified reference values)

1.3.2 Monitoring the procedure

The use of quality control cards which can be filled in either by computer or by hand has been shown to be helpful in testing, monitoring and recording the most important control parameters. Warning, control and intervention levels are distinguished. Warning levels are thresholds which may be exceeded once, while immediate action must be taken if control or intervention levels are exceeded. In an initial phase, one or more control samples are analysed with each analytical series. For each control sample the precision and, if possible, accuracy of the analyses is determined (comparison with expected values). If the quality of the analysis is acceptable, the warning and intervention levels are defined and the quality control cards drawn up. In the subsequent control period, control samples are investigated to monitor whether the defined levels are being observed. The results from these control analyses or the characteristic statistical data derived from them are continually noted in the quality control cards. The x-axis represents the serial numbers (or working days); the control values are entered on the y-axis. Changes in the control values can be seen directly; the control card can be evaluated visually immediately after each new entry. The quality control card therefore allows rapid recognition of errors; prescribed control criteria indicate an "out-of-control situation" for the monitored analytical procedure.

1.3.3 Monitoring the procedure with quality control cards [8]

Principle:

Optical representation of the quality on the basis of the
- Quality target
- Quality thresholds
  (Warning and control or intervention levels are distinguished)

Steps required for the introduction of a quality control card:

- In the initial phase one or more control samples for each analytical series is analysed.
- For each control sample the precision and, if possible, accuracy of the analytical results is then determined. If the quality of the analysis is acceptable, warning and intervention levels are defined and the quality control cards drawn up.
- In the control period, observance of the defined levels is monitored by the investigation of further control samples.

With one quality control card, one item from the following list of statistical quality data can be monitored.
- Individual value and mean value
- Recovery
- Standard deviation
- Range
- Sensitivity

1.3.4 Shewhart control card

The Shewhart control cards have proved useful for recording control parameters. They may be used, however, only for results which are normally distributed.

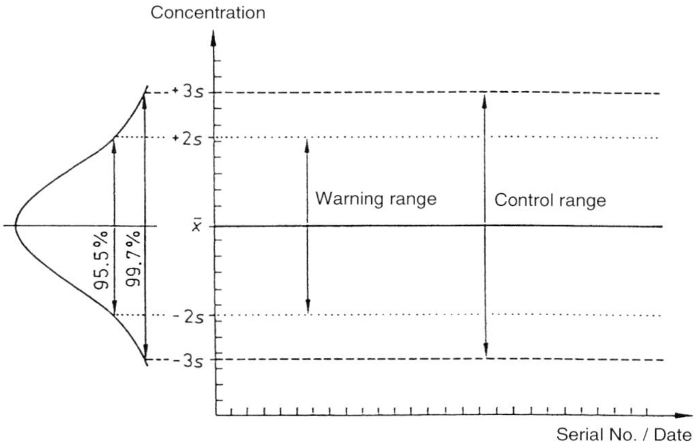

Fig. 1. Example of the construction of a Shewart control card.
Definition of "out-of-control situations" which require intervention.

1. 1 value outside of the control limits
2. 7 consecutive values on one side of the central line
3. 7 consecutive values with increasing tendency
4. 7 consecutive values with decreasing tendency
5. 2 of 3 consecutive values outside of the warning range
6. 10 of 11 consecutive values on one side of the central line

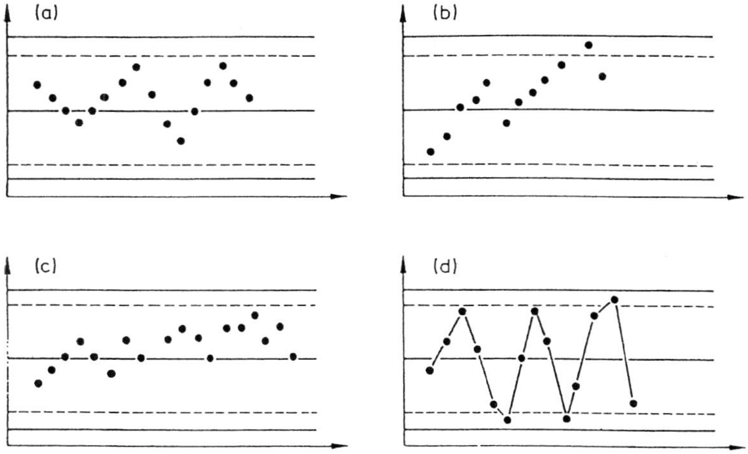

Fig. 2. Unusual entries in the Shewhart card which require investigation. (a) Cyclical changes (reasons: staff rotation, "Monday phenomenon" etc.); (b) Shift in the mean value (reasons: physical changes in the apparatus, new reagents, new devices, disposable articles etc.); (c) Trend (reasons: apparatus effects, aging of reagents etc.); (d) Many entries close to the control levels (S increased).

1.3.5 Control cards for some control parameters

Below are examples of control cards for some control parameters.

Mean value control card

Use:
- Serves to test the accuracy of the analytical procedure
- Accuracy can be monitored using certified standards

Control samples:
- Pure standard solution
- Synthetic sample
- Reference samples
- Certified reference standard

Out-of-control situations:
- As already described for the general Shewhart control card

Analytical requirements:
- at least one control sample per analytical series

Blank value control card

Use:
- Special use of the mean value control card
- Indicates the reagent and analytical system used
- The limits of the blank value control cards are defined according to the standard deviations of the mean values:

$$s_{\bar{x}} = \frac{s}{n} \tag{8}$$

Control samples:
- Reagent blank samples
- Blank samples

Out-of-control situations:
- Determination of the control and warning levels, and of the "out-of-control situations" is carried out as for the mean value control card

Analytical requirements:
- Determination of two blank values per analytical series

Recovery control cards

Use:
- Analytical procedure can be checked for matrix effects
- Generally only proportional systematic matrix influences are detected

Control samples:
- Doped real samples

Determination of the doped amount/concentration:
- The added concentration should correspond to at least the analysed value of the undoped sample
- The analysed value of the doped sample may not exceed the upper working range of the analytical procedure
- The added substance should have
  - comparable solubility
  - comparable reactivity
  - comparable oxidation state etc.

Construction of the recovery control card:
- Corresponds in design and decision criteria to the mean value control card
- After an initial phase of at least 20 determinations, the mean recovery (MR) and the standard deviation are calculated

$$MR = \frac{(x_a - x_o)}{x_{add}} \qquad (9)$$

x_a is the result of the analysis of the doped sample
x_o is the result of the analysis of the original sample
x_{add} is the added standard concentration

Out-of-control situations:
- To evaluate the recovery control cards, the out-of-control rules of the Shewhart control cards are used

Analytical requirements:
- A recovery control card should be filled in for each type of matrix
- One real control sample daily per sample matrix (concentration of the sample approximately in the middle of the concentration range)

Range control card (for groups of $n < 10$)

Use:
- Serves to check precision
- The range is the difference between the highest and lowest individual result with repeated analyses (in the analytical laboratory usually two determinations)
- Should only be used if the control limits are not set by external requirements

Control samples, analytical requirements:
- Repeated determinations of real samples should be carried out at the beginning and end of an analytical series
- As the range and the standard deviations are often concentration-dependent, the real samples should be within one concentration range, or the relative range in percent (*range/mean* × 100) must be used.

The following parameters must be determined:
- The number of repeated analyses per subgroup (n_i)

- The number (N) of subgroups
- The range (R_i) of subgroup i
- The mean value (R) of the ranges
- The variance of the total analysis (s^2)
- The lower and upper warning and control levels

Calculation of the control levels:

$$CL = D_{cl} \cdot R$$

$$CU = D_{cu} \cdot R$$

$$WL = D_{wl} \cdot R$$

$$WU = D_{wu} \cdot R$$

CL is the lower control level
CU is the upper control level
WL is the lower warning level
WU is the upper warning level
D_{cl} is the statistical factor for the lower control level
D_{cu} is the statistical factor for the upper control level
D_{wl} is the statistical factor for the lower warning level
D_{wu} is the statistical factor for the upper warning level
R is the mean value of the ranges

Out-of-control situations:
- A range is above the upper control level
- A range is below the lower control level (only applies if CL >0)
- Seven consecutive values have an increasing or decreasing tendency
- Seven consecutive values are above the mean range

Standard deviation control card (for groups of $n > 10$)

Use:
- Monitoring of the range of scatter of all the values
- Can reveal systematic errors

Construction of a standard deviation control card:
- Central line is at the level of the mean standard deviation
- Control levels calculated from the X^2 distribution

Analytical requirements:
- Can be used best with groups of $n > 10$

1.4 Quality control of data collecting and processing systems

When using computer-aided analytical procedures, the reliability of the analytical results depends also on the monitoring and validation of the computerised systems.
The general principles are described below. Details can be found in the OECD document "The application of the principles of GLP to computerised systems" [9].
Computerised systems used for monitoring should be constructed for the purpose, have sufficient capacity and be suitable for the use intended. Suitable procedures for testing and servicing these systems should be available, and the systems should be developed, validated and run in accordance with defined principles (e. g. GLP).
Demonstration that a computerised system is suitable for the intended use is described as computer validation.
The validation process more or less ensures that a computerised system meets the prescribed specifications. Validation should be carried out using a formal validation plan before the system is used for monitoring.
The operating procedures (SOPs) drawn up should include statements about the responsibilities of the co-ordinator, test organiser, personnel and quality control. They must contain statements about qualifications and training, apparatus (site etc.) and equipment (hardware and software), servicing and restoring functionality after system failures.
Particular care should be taken in the treatment and archiving of data. If computerised systems are used for collecting and processing raw data, creating reports or storing raw data, the system should be designed in such a way as to allow the generation and storage of a complete audit trail at all times, so that all changes in the data may be reproduced without the original data being made unrecognisable. By using (electronic) signatures which also give the date and time, it should be possible to trace all changes in the data back to the persons who made these changes. Reasons for the changes should be given. When raw data are stored, it is necessary to provide for long-term storage.
Information affecting the monitoring process such as servicing and calibration data, which are necessary for confirming the validity of the raw data or for the reconstruction of a procedure or a test, must be stored in the archives.
Documented procedures for the safety and protection of hardware, software and data against falsification, unauthorised amendment or loss should be available.

1.4.1 Validation of computerised systems

The following points must be taken into consideration.

Acceptance

Sufficient documentation should be available to confirm that the system was developed in a verifiable way and preferably according to recognised quality and technical standards (such as ISO/9001). In addition it must be certified that the system has been checked by the analytical laboratory to see that it fulfils the acceptance criteria. The formal acceptance test involves the carrying out of the necessary tests and the storage of the documentation for all test procedures, test data and test results.

If the systems are obtained from a supplier, most of the documentation drawn up during development remains with the manufacturer. In this case, certification of a formal estimation of the reliability and/or testing of the method of working of the manufacturer should be available in the analytical laboratory.

Procedure for monitored system change

The procedure of monitored system change for hardware and software consists of the formal authorisation to carry out the change and the documentation of each change to a computerised system during use. A procedure for monitored system change is necessary when an intended change to the system could influence its valid functioning.
The procedure should describe the method of evaluation with which the necessary extent of retesting the system to ensure its valid function is to be determined.
Regardless of the reason for the change (external manufacturer or own system), sufficient information is part of the procedure of monitored system change. The procedure of monitored system change must guarantee data integrity.

Supportive measures

Supportive measures should be available which guarantee that the computerised system functions correctly and is used properly. Supportive measures can, for example, include system management, training, servicing, technical support, monitoring and/or system evaluation. Evaluation of the system's performance involves the formal testing of a system at regular intervals to check that the prescribed performance criteria, such as reliability, response behaviour, capacity are fulfilled.

1.4.2 Documentation

The points listed below describe the minimum necessary documentation of the development, validation, operation and servicing of computerised systems.

Guidelines

Guidelines should be available for the nature, requirements, conception, validation, tests, installation, operation, servicing, staff, process testing and single checks, monitoring and decommissioning of computerised systems.

Description of the user software

The documentation necessary for all applications includes:
- The name of the user software or its identification code and a detailed and comprehensible description of its purpose
- Hardware (with model number), on which the user software runs
- Operating system and other software
- Programming language and/or databanks used
- Main functions of the application
- Description of the type of data and data flow

- File structures, error and alarm messages, and algorithms
- Components of the user software with version numbers
- Configuration and communication links

Standard operating procedures (SOPs)

Most of the documentation describing the use of the computerised system, should be in the form of SOPs.

1.4.3 Archiving

The principles for archiving data must be uniformly implemented for all types of data. It is therefore necessary that, for the storage of electronic data, equivalent procedures are introduced for checking access, indexing and retrieving, and recovery after failures. Electronically stored data and the corresponding documentation may not be destroyed without authorisation from the head of the analytical laboratory. Data giving additional information on the computerised system, such as source code and details about developments, validation, operation, servicing and monitoring, should be stored for at least as long as the analytical data.

1.5 References

[1] *Deutsches Institut für Normung e.V.* (DIN) (1987). DIN 55350-13 Begriffe der Qualitätssicherung und Statistik; Begriffe zur Genauigkeit von Ermittlungsverfahren und Ermittlungsergebnissen. Beuth Verlag, Berlin
[2] *Europäisches Komitee für Normung* (CEN) (1994). DIN EN 482 Arbeitsplatzatmosphäre-Allgemeine Anforderungen an Verfahren zur Messung von chemischen Arbeitstoffen. Brüssel. Beuth Verlag, Berlin
[3] *International Organization for Standardization* (ISO) (1994). ISO 5725-1 Genauigkeit (Richtigkeit und Präzision) von Meßverfahren und Meßergebnissen – Teil 1: Begriffe und allgemeine Grundlagen. Geneva, Beuth Verlag, Berlin
[4] *Länderausschuß für Arbeitsschutz und Sicherheitstechnik* (LASI) (1996). LV 4 Qualitätssicherungs-Handbuch (QSH)
[5] *International Organization for Standardization* (ISO) (1997). ISO Guide 32 Calibration in analytical chemistry and use of certified reference materials, Geneva. Beuth Verlag, Berlin.
[6] *International Organization for Standardization* (ISO) (1989). ISO Guide 35 Certification of reference materials – General and statistical principles, Geneva. Beuth Verlag, Berlin.
[7] *Institute for Reference Materials and Measurements* (IRMM) (1998). BCR Reference Materials 1998, Geel
[8] *Funk W, Dammann V, Donnevert G* (1992) Qualitätssicherung in der Analytischen Chemie. VCH Verlag, Weinheim
[9] *Organisation for Economic Co-operation and Development* (OECD) (1995) The application of the principles of GLP to computerised systems. Environment monograph No. 116. In: OECD series on principles of good laboratory practice and compliance monitoring, No 10, Paris

Author: *W. Riepe, M.R. Lahaniatis*

Analytical Methods

Azinphos-methyl

Method number	1
Application	Air analysis
Analytical principle	High performance liquid chromatography
Completed in	September 1997

Summary

To determine the levels of azinphos-methyl in the air, measured air volumes are drawn through Tenax adsorption tubes by a battery-operated pump. The adsorbed substance is extracted with acetonitrile and determined using a UV detector after liquid chromatographic separation. Quantitative evaluation is carried out using a calibration curve. The peak areas obtained using calibration standards are plotted against the azinphos-methyl concentrations.

Precision:	Standard deviation (rel.): s = 1.6%
	Mean variation: u = 4.0%
	for 5 activated carbon tubes each loaded with 26 µg azinphos-methyl
Limit of quantification:	25 µg/m^3 azinphos-methyl for a sampled air volume of 720 L
Recovery:	$\bar{\eta}$ = 98.3 or 98.7% at concentrations of 26 or 300 µg/m^3
Sampling recommendation:	Sampling time: 6 hours
	Sampled air volume: 720 L

Azinphos-methyl [CAS No. 86-50-0]

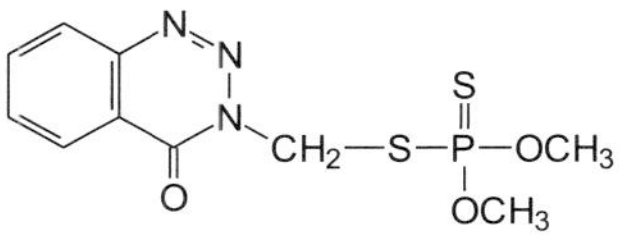

Azinphos-methyl, a non-systemic insecticide introduced in 1955, has a broad spectrum of action and may be applied to numerous crops.

Azinphos-methyl is a colourless, crystalline solid with a melting point of 73 °C and a molecular weight of 317.1.
Azinphos-methyl has a MAK value of 0.2 mg/m^3 (2001) – measured as the inhalable aerosol fraction. Because of the danger of absorption through the skin, azinphos-methyl has been marked with an “H” in the List of MAK and BAT Values [1].

Author: *K. Riegner*
Examiner: *W. Kleiböhmer*

Azinphos-methyl

Method number 1

Application Air analysis

Analytical principle High performance liquid chromatography

Completed in September 1997

Contents

1 General principles

To determine the levels of azinphos-methyl in the air measured air volumes are drawn through Tenax adsorption tubes by a battery-operated pump. The adsorbed substance is extracted with acetonitrile and determined with a UV detector after liquid chromatographic separation. Quantitative evaluation is carried out using a calibration curve. The peak areas of the calibration standards are plotted against the azinphos-methyl concentrations.

2 Equipment, chemicals and solutions

2.1 Equipment

Adsorption tube type NIOSH, Tenax® (e.g. Günther Karl OHG, Gau-Algesheim, Germany Order No. GK-26-35-03-GO)
Sampling pump (e.g. HI FLOW Sampler, model HFS 113 A, Gilian Instruments, New Jersey, USA)
Gasmeter
Thermometer
Barometer
High performance liquid chromatograph with gradient pump system and UV detector (220 nm)
Glass cutter
10 mL and 25 mL volumetric flasks
50 µL, 100 µL and 250 µL precision syringes
20 mL flanged vials with Teflon-coated butyl rubber stoppers
Laboratory shaker

2.2 Chemicals

Azinphos-methyl (e.g. Ehrenstorfer, Augsburg, Germany)
Acetonitrile, suitable for HPLC gradient elution
Water, suitable for HPLC

2.3 Solutions

Stock solution: To prepare the stock solution, 25 mg azinphos-methyl is weighed exactly into a 25 mL volumetric flask. Then the flask is filled to the mark with acetonitrile (1 g/L).

Elution solution: To prepare the elution solution 1 mL of the stock solution is pipetted into a 10 mL volumetric flask containing approximately 5 mL acetonitrile. Then the flask is filled to the mark with acetonitrile (100 mg/L).

Calibration standards: From this elution solution calibration standards containing 0.1–10 mg/L of azinphos-methyl are prepared by diluting with acetonitrile. The following pipetting scheme is used (Table 1):

Table 1. Pipetting scheme for the preparation of the calibration standards.

Volume of the elution solution µL	Final volume of the calibration standards mL	Concentration of the calibration standards mg/L
10	10	0.1
100	10	1.0
200	10	2.0
500	10	5.0
1000	10	10.0

3 Sample collection and preparation

The adsorption tube is opened at both ends and connected to the inlet of the pump with a short tube. The adsorption tube contains two adsorption phases separated by cotton wool; the larger phase (100 mg Tenax) is closest to the inlet of the tube during air sampling. The second, smaller adsorption phase (50 mg Tenax) serves to check for possible breakthrough of the substance during sampling. For sampling, air is drawn through the adsorption tube at a flow rate of 2 L/min for six hours. After sampling, the air sample volume is noted. In addition, the parameters important for determining the concentration such as temperature and air pressure at the site of sampling must be determined. The closed tubes should be stored in a refrigerator until sample preparation is performed.
For preparation, the collection phase and the control phase are each transferred separately to 20 mL flanged vials; the cotton wool separating the two phases and the upper piece of cotton wool at the tube inlet are analysed together with the collection phase.
10 mL acetonitrile (collection phase) and 5 mL acetonitrile (control phase) are added to the flanged vials and the vials are closed. The substance is extracted from the adsorption material by shaking the tube on a laboratory shaker for 10 minutes. 20 µL of the acetonitrile solution is then injected without further processing into the high performance liquid chromatograph for separation.
In each analysis series a reagent blank is prepared and analysed in the same way.

4 Operating conditions for high performance liquid chromatography

Precolumn:	Material:	RP-18 LiChrospher 100
	Length:	4 mm
	Internal diameter:	4 mm
	Particle size:	5 µm

Column:	Material:	RP-18 LiChrospher 100
	Length:	25 cm
	Internal diameter:	4 mm
	Particle size:	5 µm
Column temperature:	40 °C	
Solvent:	A = water	
	B = acetonitrile	
Gradient:	0 minutes	45% B
	2 minutes	45% B
	8 minutes	70% B
	10 minutes	70% B
	11 minutes	45% B
	14 minutes	45% B
Flow rate:	1.0 mL/min	
Detector:	UV detector	
Detection wavelength:	220 nm	
Injection volume:	20 µL	

Under the high performance liquid chromatographic conditions described above, the retention time was about 9.3 minutes (see Figure 1).

5 Analytical determination

Under the conditions described above, 20 µL of the acetonitrile solution is injected into the high performance liquid chromatograph using an injection loop. After high performance liquid chromatographic separation of the azinphos-methyl from the other components collected on the Tenax and extracted with acetonitrile, the azinphos-methyl is detected by the UV detector.

6 Calibration

20 µL of each of the calibration standards (see Section 2.3) is injected into the high performance liquid chromatograph and detected by the UV detector. To draw the calibration curve, the measured peak areas or heights are corrected by subtraction of the reagent blank values and then plotted against the azinphos-methyl concentrations used in mg/L (see Figure 2).
Each solution is analysed twice, and the mean value is used in the calculation.

7 Calculation of the analytical result

Using the peak areas or heights obtained after subtraction of the reagent blank values, the azinphos-methyl concentration in mg/L acetonitrile is read from the calibration curve. If analysis of the control phase reveals an azinphos-methyl concentration of more than 10% of the total azinphos-methyl concentration (from the collection phase and control phase), a breakthrough has occur. In this case sampling must be repeated under different conditions (e. g. a lower air sample volume).
The concentration by weight ρ (mg of azinphos-methyl per m^3 of air) in the sample air is calculated according to the following equation:

$$\rho = \frac{a \cdot b}{V_Z \cdot \eta} \tag{1}$$

At 20 °C and 1013 hPa:

$$\rho_0 = \rho \cdot \frac{273 + t_a}{293} \cdot \frac{1013}{p_a} \tag{2}$$

The corresponding concentration by volume – independent of the variables pressure and temperature – is given by:

$$\sigma = \rho_0 \cdot \frac{V_m}{M} = \rho \cdot \frac{273 + t_a}{293} \cdot \frac{1013}{p_a} \cdot \frac{24.1}{317.1} \tag{3}$$

$$\sigma = \rho \cdot \frac{273 + t_a}{p_a} \cdot 0.263$$

For $t_a = 20$ °C and $p_a = 1013$ hPa:

$$\sigma = \rho \cdot 0.076 \ \frac{\text{mL}}{\text{mg}} \tag{4}$$

where:

- a is the azinphos-methyl concentration in acetonitrile in mg/L taken from the calibration curve
- b is the volume of acetonitrile used for extraction in L
- V_z is the air sample volume in m^3
- η is the recovery
- t_a is the temperature of the ambient air in °C
- p_a is the air pressure of the ambient air in hPa
- ρ is the azinphos-methyl concentration in the ambient air in mg/m^3 at t_a and p_a
- ρ_c is the azinphos-methyl concentration in the ambient air in mg/m^3 at 20 °C and 1013 hPa

M is the molecular weight of azinphos-methyl
V_m is the molecular volume in L/mole
σ is the azinphos-methyl concentration in the ambient air in mL/m^3

8 Reliability of the method

8.1 Precision

To determine the precision, 5 Tenax adsorption tubes were each loaded with 26 µg azinphos-methyl in dissolved form and subjected to the preparation procedure described above. The standard deviation (rel.) was $s = 1.6\%$ and the mean variation $u = 4.0\%$.

8.2 Recovery rate

To determine the recovery, defined amounts of the substance in dissolved form were added to each of 5 Tenax adsorption tubes. Care was taken that the solution was added to the middle of the tube and that the spiked volume did not exceed 50 µL. The solvent was removed by drawing air through the tube (2 L/min) for about ten minutes. The tubes were then prepared and analysed as described in Sections 3 and 4. Table 2 shows the resulting recovery values.

Table 2. Recovery η.

Concentration mg/m^3	Recovery η %	Mean recovery: $\bar{\eta}$ %
0.026	96.5–101	98.3
0.300	97.1–102	98.7

8.3 Quantification limit

Under the given analytical conditions the quantification limit of the procedure is 25 $\mu g/m^3$.
By a slight modification of the method, described below, the quantification limit can be lowered to 0.5 $\mu g/m^3$. During sample preparation the collection and control phases are mixed separately each with 3 mL acetonitrile and the substance is extracted from the adsorption material by shaking the tubes for 30 minutes on a laboratory shaker. Before liquid chromatographic separation 30% water is added to the extract. 200 µL of the acetonitrile/water solution is then injected directly into the high performance liquid chromatograph.

8.4 Shelf-life

To check the shelf-life, Tenax tubes were spiked with defined amounts of azinphos-methyl. The spiked amount was equivalent to about 0.05 mg/m^3 for an air sample volume of 720 L. The solvent was removed by drawing air through the tube (2 L/min) for about ten minutes. The tubes were then stored in either a refrigerator (approx. +1 to +6 °C) or a deep-freeze (approx. –22 to –25 °C). After storage for 11 days the adsorption tubes were prepared and analysed as described in Sections 3 and 4 in the form of a repeated determination. Under both kinds of storage conditions recovery was >99%.

9 Discussion of the method

The method described for determining azinphos-methyl in air consist of two steps. In the first step the substances in the air are deposited in the collection system and in a second step extracted from it. Tenax has proved to be an ideal collection phase. Tenax adsorption tubes are also suitable for collecting gaseous and particle-bound residues of other pesticides from the air [2]. Care must be taken when determining the recovery that the solution is added to the middle of the tube and that the spiked volume is 10–50 µL. Smaller volumes increase the inaccuracy of the dose, larger volumes lead, however, to increased substance losses during removal of the solvent. Recovery was found to be adequate with "worst case" climatic conditions (T = 35 °C; relative humidity = 80%).

Instruments used:
Liquid chromatograph HP 1050 with UV detector and autosampler

10 References

[1] *Deutsche Forschungsgemeinschaft.* MAK- und BAT-Werte-Liste 2001. Senatskommission zur Prüfung gesundheitsschädlicher Arbeitsstoffe, Mitteilung 37. WILEY-VCH Verlag, Weinheim

[2] *Riegner K, Schmitz JR* (1994). Erzeugung einer Testatmosphäre und Abscheidung gasförmiger und partikelgebundener Pflanzenschutzmittelrückstände aus Luft auf Tenax®-Sammelröhrchen. Pflanzenschutz-Nachrichten Bayer 47: 161–176

Author: *K. Riegner*
Examiner: *W. Kleiböhmer*

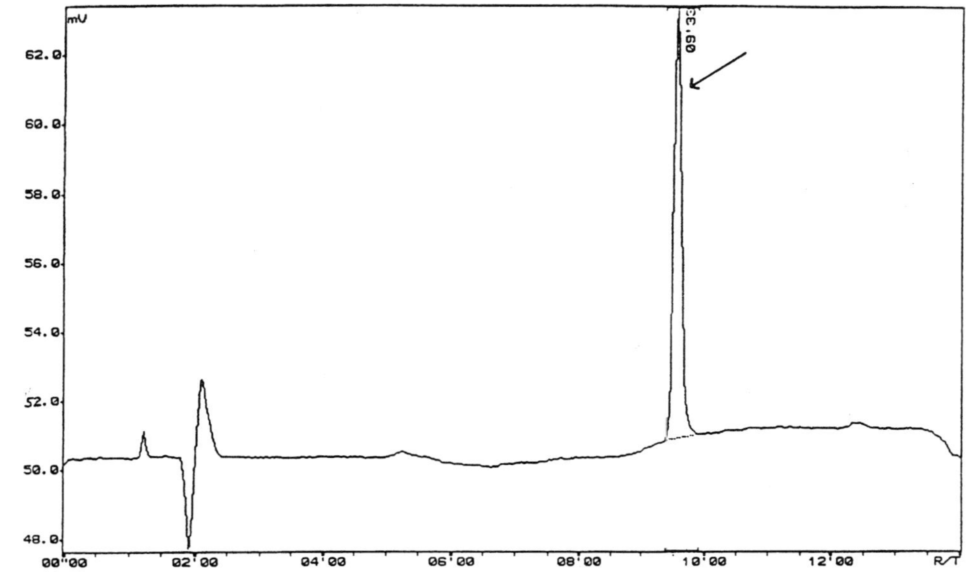

Fig. 1. HPLC chromatogram of a standard spiked with 2 mg azinphos-methyl per litre.

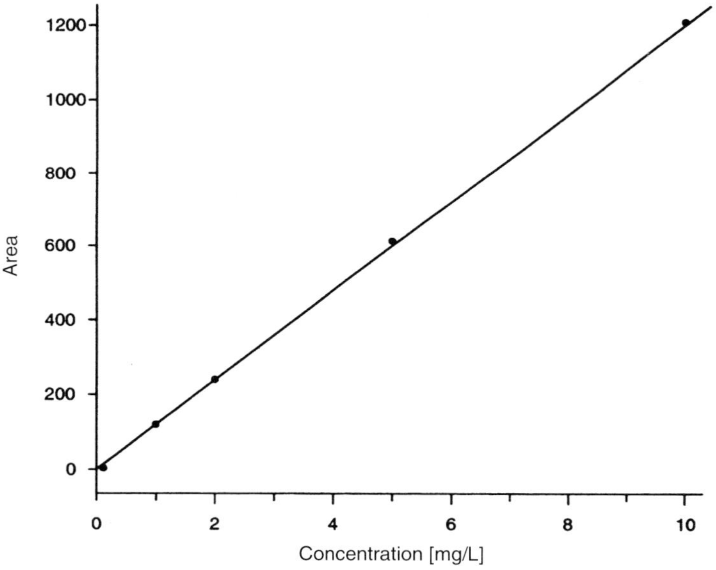

Fig. 2. Example of a calibration curve.

Fenthion

Method number 1

Application Air analysis

Analytical principle Gas chromatography

Completed in November 1998

Summary

To determine the levels of fenthion in the air, measured air volumes are drawn through Tenax adsorption tubes by a battery-operated pump. The adsorbed substance is extracted with *n*-butyl acetate and determined with an MSD detector after gas chromatographic separation. Quantitative evaluation is carried out using a calibration curve. The peak areas obtained using calibration standards are plotted against the fenthion concentrations.

Precision:	Standard deviation (rel.)	s = 1.7 or 5.0%
	Mean variation	u = 5.4 or 15.9%
	for the fenthion concentrations 1.8 and 200 µg/m^3 and n = 4 determinations	
Limit of quantification:	1.8 µg/m^3 fenthion with a sampled air volume of 720 L	
Recovery rate:	$\bar{\eta}$ = 103 or 95.6% at concentrations of 1.8 or 200 µg/m^3	
Sampling recommendation:	Sampling time:	6 hours
	Sampled air volume:	720 L

Fenthion [CAS No. 55-38-9]

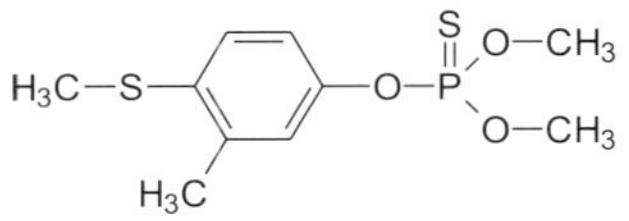

Fenthion is an insecticide against biting and sucking insects, in particular fruit flies and domestic pests, introduced by BAYER in 1958.

Fenthion is a colourless, oily liquid with a boiling point of 87 °C at 0.014 hPa, a molecular weight of 278.3 and a vapour pressure of 3.6×10^{-6} hPa at 20 °C.
Fenthion has a MAK value of 0.2 mg/m^3 (2001) – measured as the inhalable aerosol fraction. Because of the danger of absorption through the skin, fenthion has been marked with an "H" in the List of MAK and BAT Values [1].

Author: *K. Riegner*
Examiner: *W. Kleiböhmer*

Fenthion

Method number 1

Application Air analysis

Analytical principle Gas chromatography

Completed in November 1998

Contents

1 General principles

To determine the levels of fenthion in the air measured air volumes are drawn through Tenax adsorption tubes by a battery-operated pump. The adsorbed substance is extracted with *n*-butyl acetate and determined with an MSD detector after gas chromatographic separation. Quantitative evaluation is carried out using a calibration curve. The peak areas of the calibration standards are plotted against the fenthion concentrations.

2 Equipment, chemicals and solutions

2.1 Equipment

Adsorption tube type NIOSH, Tenax® (e.g. Günther Karl OHG, Gau-Algesheim, Germany Order No. GK-26-35-03-GO)
Sampling pump (e.g. HI FLOW Sampler, model HFS 113 A, Gilian Instruments, New Jersey, USA)
Gasmeter
Thermometer
Barometer
Flowmeter or soap bubble meter
Gas chromatograph with mass selective detector and autosampler
Glass cutter
10 mL, 100 mL and 1000 mL volumetric flasks
10 mL flanged vials with Teflon-coated butyl rubber stoppers
Laboratory shaker

2.2 Chemicals

Fenthion, $w = 99.7\%$ (e.g. Ehrenstorfer, Augsburg, Germany)
n-Butyl acetate, analytical grade

2.3 Solutions

Stock solution:
To prepare the stock solution, 100 mg fenthion is weighed exactly into a 1000 mL volumetric flask. The flask is then filled to the mark with *n*-butyl acetate (100 mg/L).

Elution solution:
To prepare the elution solution 1 mL of the stock solution is pipetted into a 100 mL volumetric flask containing approximately 50 mL *n*-butyl acetate. The flask is then filled to the mark with *n*-butyl acetate (1 mg/L).

Calibration standards:
From this elution solution calibration standards containing 0.09–0.78 mg/L of fenthion are prepared by diluting with *n*-butyl acetate. The following pipetting scheme is used:

Table 1. Pipetting scheme for the preparation of the calibration standards.

Volume of the elution solution mL	Final volume of the calibration standards mL	Concentration of the calibration standards mg/L
0.9	10	0.09
1.8	10	0.18
2.6	10	0.26
5.2	10	0.52
7.8	10	0.78

3 Sample collection and preparation

The adsorption tube is opened at both ends and connected to the inlet of the pump by a short tube. The adsorption tube contains two adsorption phases separated by cotton wool; the larger phase (100 mg Tenax) is closest to the inlet of the tube during air sampling. The second, smaller adsorption phase (50 mg Tenax) serves to check for breakthrough of the substance during sampling. For sampling, air is drawn through the adsorption tube at a flow rate of 2 L/min for six hours (720 L). After sampling, the air sample volume is noted. The temperatures and atmospheric pressures measured during sampling are noted. The closed tubes should be stored in a refrigerator until sample preparation is performed.

For preparation, the collection phase and the control phase are each transferred separately to 10 mL flanged vials; the cotton wool separating the two phases and the upper piece of cotton wool at the tube inlet are analysed together with the collection phase.

3 mL n-butyl acetate (collection phase) and 1.5 mL n-butyl acetate (control phase) are added to the flanged vials and the vials are closed. The substance is extracted from the adsorption material by shaking the tube on a laboratory shaker for 10 minutes. The extraction procedures are repeated twice. The extracts are combined (after adding *n*-butyl acetate the final volumes are 10 mL for the collection phase and 5 mL for the control phase). 1 µL of the *n*-butyl acetate solution is then injected without further processing into the gas chromatograph for separation. If necessary the solution must first be diluted with *n*-butyl acetate.

In each analysis series a reagent blank is prepared and analysed in the same way.

4 Operating conditions for gas chromatography with mass selective detector

Column:	Material:	Quartz capillary
	Internal diameter:	0.20 mm
	Length:	12 m
	Stationary phase:	HP 5
	Film thickness:	0.33 µm
Temperatures:	Furnace:	100 °C; 1 min
		20 °C/min up to 240 °C
		240 °C; 2 min
	Injector:	300 °C
	Transfer line to ion source:	305 °C
Injection volume:	1 µL	
Sample injection:	Splitless	
	plitless time:	0–1 minutes
Carrier gas:	Helium (60 kPa column pressure)	
Carrier gas flow rate:	30 mL/min	
Detector:	Quadrupol MS	
Detector mode:	SIM	
Dwell time (recording time/ion):	75 ms	

Under the gas chromatographic conditions described above, the retention time was about 7.3 minutes (see Figure 1).

5 Analytical determination

Under the conditions described above, 1 µL of the *n*-butyl acetate solution is injected into the gas chromatograph. After gas chromatographic separation of the fenthion from the other components collected on the Tenax and extracted with *n*-butyl acetate, the fenthion is detected with the mass selective detector. The “Single Ion Monitoring” (SIM) mode, which allows sensitive and specific detection of selected masses, is selected. In the present case the “mole peak” (m/z = 278) serves for quantification (see Figure 2).

6 Calibration

1 µL of each of the calibration standards (see Section 2.3) is injected into the gas chromatograph and detected using the MSD detector. To draw the calibration curve, the measured peak areas or heights are corrected by subtraction of the reagent blank values and then plotted against the fenthion concentration used in mg/L. Each solution is analysed twice and the mean value is used for calculation.
The linearity of the detector has been checked and verified in the injection concentration range from 0.002–2 mg/L.

7 Calculation of the analytical result

Using the peak areas or heights obtained after subtraction of the reagent blank values, the fenthion concentration in mg/L *n*-butyl acetate is read from the calibration curve. If analysis of the control phase reveals a fenthion concentration of more than 10% of the total fenthion concentration (from the collection and control phases), breakthrough has taken place. In this case sampling must be repeated under other conditions (e.g. a lower air sample volume).
The concentration by weight ρ (mg of fenthion per m^3 of air) in the sample air is calculated according to the following equation:

$$\rho = \frac{a \cdot b}{V_Z \cdot \eta} \qquad (1)$$

At 20 °C and 1013 hPa:

$$\rho_0 = \rho \cdot \frac{273 + t_a}{293} \cdot \frac{1013}{p_a} \qquad (2)$$

The corresponding concentration by volume – independent of the variables pressure and temperature – is given by:

$$\sigma = \rho_0 \cdot \frac{V_m}{M} = \rho \cdot \frac{273 + t_a}{293} \cdot \frac{1013}{p_a} \cdot \frac{24.1}{278.3} \qquad (3)$$

$$\sigma = \rho \cdot \frac{273 + t_a}{p_a} \cdot 0.299$$

For t_a = 20 °C and p_a = 1013 hPa:

$$\sigma = \rho \cdot 0.086 \, \frac{\text{mL}}{\text{mg}} \qquad (4)$$

where:

a	is the fenthion concentration in n-butyl acetate in mg/L taken from the calibration curve
b	is the volume of n-butyl acetate used for extraction in L
V_z	is the air sample volume in m^3
η	is the recovery
t_a	is the temperature of the ambient air in °C
p_a	is the air pressure of the ambient air in hPa
ρ	is the fenthion concentration in the ambient air in mg/m^3 at t_a and p_a
ρ_o	is the fenthion concentration in the ambient air in mg/m^3 at 20 °C and 1013 hPa
M	is the molecular weight of fenthion
V_m	is the molecular volume in L/mole
σ	is the fenthion concentration in the ambient air in mL/m^3

8 Reliability of the method

8.1 Precision

The precision was calculated by determining the recovery. In 4 assays for each concentration, the standard deviations (relative) s and mean variations u listed in Table 2 were found.

Table 2. Standard deviation (relative) s and mean variation u, n = 4 determinations.

Concentration µg/m^3	Standard deviation (relative) s %	Mean variation u %
1.8	1.7	5.4
200	5.0	15.9

8.2 Recovery rate

To determine the recovery, defined amounts of the substance in dissolved form were added to each of 4 Tenax adsorption tubes. The solvent was removed by drawing air through the tube (2 L/min) for about ten minutes. The samples were then prepared and analysed as described in Sections 3 and 4.
Table 3 shows the resulting recovery values.

Table 3. Recovery η.

Concentration mg/m^3	Recovery η %	Mean recovery $\bar{\eta}$ %
1.8	101–105	103
200	91.3–101	96

The substance was not detected on the second phase of the Tenax adsorption tube. The breakthrough therefore represents less than 0.1% of the amount of substance used of 200 µg/m$^{3.}$

8.3 Quantification limit

Under the given analytical conditions the quantification limit of the procedure is 1.8 µg/m^3.

8.4 Shelf-life

To check the shelf-life of adsorption tubes loaded with fenthion, Tenax tubes were spiked with defined amounts of the substance. The spiked amount was equivalent to about 4 µg/m^3 for an air sample volume of 720 L. After air was drawn through the tube for ten minutes (to remove the solvent) the tubes were stored in either refrigerator (<8 °C) or a deep-freeze (<–18 °C). After the given number of days in storage, the adsorption tubes were prepared and analysed as described in Sections 3 and 4 in the form of a repeated determination.

Table 4. Checking the shelf-life.

Number of days	Temperature °C	Recovery %
3	<8	99.4
3	<–18	97.3
8	<8	99.9
8	<–18	98.7

9 Discussion of the method

The method permits the determination of fenthion in the air in a concentration range of 1.8 μg/m^3 to 200 μg/m^3. The procedure has two steps. In the first step the substances in the air (gaseous components and particles) are deposited in the collection system, and in a second step extracted from it. Tenax has proved to be an ideal collection phase. Tenax adsorption tubes are suitable for the determination of fenthion in the air, both as particles and in gaseous form [2]. Care must be taken when determining the recovery that the solution is added to the middle of the tube and that the spiked volume is 10–50 μL. Smaller volumes increase the inaccuracy of the dose, larger volumes lead, however, to increased substance losses during removal of the solvent. Recovery was found to be adequate with "worst case" climatic conditions (T = 35 °C; humidity (relative) = 80%) [2].

Instruments used: Gas chromatograph HP 5790 with MSD HP 5790 and autosampler HP 7673 a

10 References

[1] *Deutsche Forschungsgemeinschaft* (2001) MAK- und BAT-Werte-Liste 2001. Senatskommission zur Prüfung gesundheitsschädlicher Arbeitsstoffe, Mitteilung 37. WILEY-VCH Verlag, Weinheim

[2] *Riegner K, Schmitz JR* (1994). Erzeugung einer Testatmosphäre und Abscheidung gasförmiger und partikelgebundener Pflanzenschutzmittelrückstände aus Luft auf Tenax®-Sammelröhrchen. Pflanzenschutz-Nachrichten Bayer 47: 161–176

Author: *K. Riegner*
Examiner: *W. Kleiböhmer*

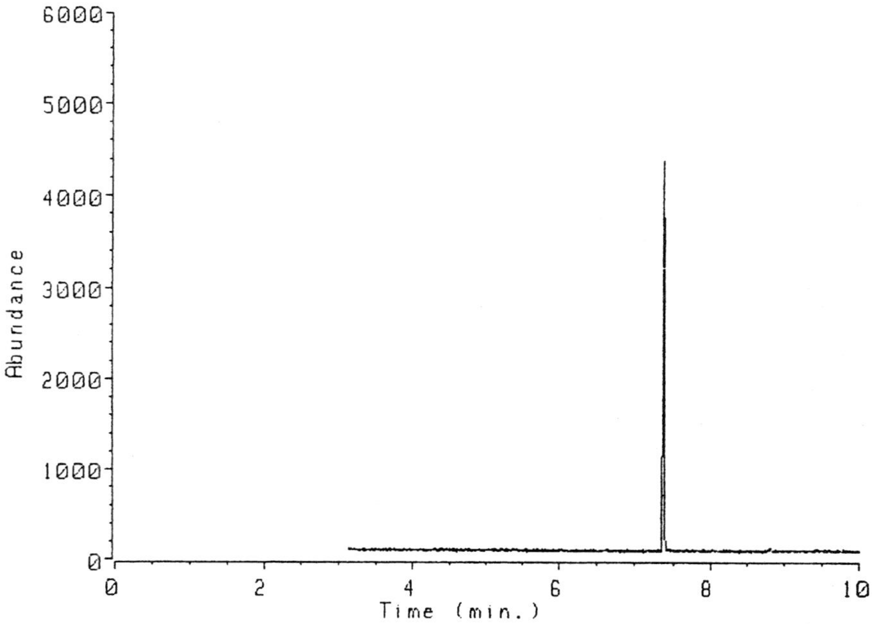

Fig. 1. Gas chromatogram of a standard spiked with 0.144 mg fenthion per litre.

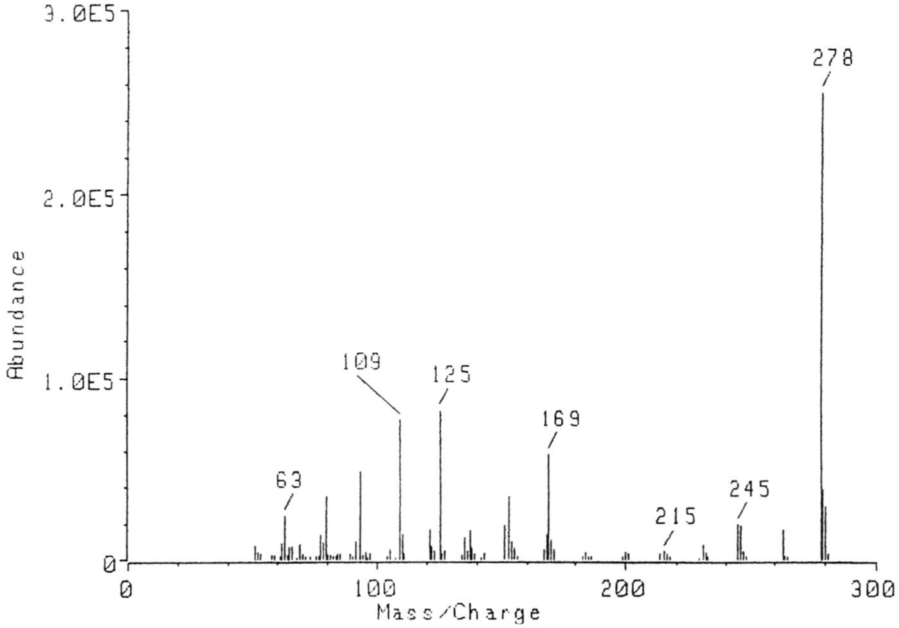

Fig. 2. Mass spectrum of fenthion.

Inorganic acid mists (H_2SO_4, H_3PO_4)

Method number 1

Application Air analysis

Analytical principle Ion chromatography

Completed in April 1997

Summary

The method described permits the determination of the mists of the inorganic acids phosphoric acid and sulfuric acid, the latter in the range from 0.1 to 2 times the currently valid threshold limit values in air [1, 2].
During sampling, the air containing the acids is drawn through a quartz fibre filter to collect the acids. Desorption is carried out with an aqueous sodium bicarbonate/sodium hydrogen carbonate solution. Quantitative determination is carried out by ion chromatography.

Quantification limit: H_2SO_4 absolute 50 ng = 0.01 mg/m^3
H_3PO_4 absolute 50 ng = 0.01 mg/m^3
for a sampled air volume of 420 L.

Recovery: H_2SO_4 $\eta = 0.97–1.00$ (97–100%)
H_3PO_4 $\eta = 0.97–1.00$ (97–100%)
for the whole range.

Precision:

Table D 1. Standard deviation (rel.) s and mean variation u, $n = 10$ determinations.

Substance	Concentration mg/m^3	Standard deviation (rel.) s %	Mean variation u %
H_2SO_4	0.01	2.6	5.8
	0.02	1.6	3.6
	0.08	0.5	1.1
	0.20	0.6	1.3
	1.00	0.9	2.0

Table D 1. (continued)

Substance	Concentration mg/m^3	Standard deviation (rel.) s %	Mean variation u %
H_3PO_4	0.01	3.2	7.1
	0.08	1.4	3.1
	0.20	0.7	1.6
	1.00	2.4	5.3
	2.00	2.9	6.4

Sampling recommendation:	Sampling time:	2 hours
	Sampled air volume:	420 L (at concentrations up to 3 times the threshold limit value in air)

Sulfuric acid (H_2SO_4) [CAS No. 7664-93-9]

Sulfuric acid is a colourless and odourless, highly hygroscopic and highly caustic liquid (molecular weight 98.08, freezing point 10 °C, boiling point 338 °C [98.3% H_2SO_4]). Sulfuric acid is miscible with water with the generation of much heat.
Sulfuric acid is widely used in industry (e.g. in the extraction of rock phosphates and other ores, in metal processing, in electroplating, for sulfonation, as a component of nitrating acid, as a desiccant, in lead batteries).
The currently valid threshold limit value is 1 mg/m^3, measured as the inhalable aerosol fraction [1]. The MAK commission recommends a MAK Value of 0.1 mg/m^3, measured as the inhalable aerosol fraction [2]. Sulfuric acid is classified in Peak limitation category I in the List of MAK and BAT Values and given an excursion factor of =1= according to TRGS 900.

Phosphoric acid (H_3PO_4) [CAS No. 7664-38-2]

Phosphoric acid (ortho) forms colourless rhomboid crystals which deliquesce in air (molecular weight 98.00, freezing point 42.3 °C, boiling point 213 °C). H_3PO_4 is readily miscible with water with the production of heat. The concentrated phosphoric acid ($\approx$90%) available commercially is a syrupy, caustic liquid.
Phosphoric acid is used e.g. in the production of phosphate fertilisers, phosphates, porcelain cements, as polymerisation catalyst, in metal processing e.g. as etching agent or in the production of flame retardants.

At present there is no threshold limit value for the acid in air which is valid in Germany. In the USA the TLV value is 1 mg/m^3 [3].

Author: *D. Breuer*
Examiners: *R. Hebisch, K. Macho*

Inorganic acid mists (H_2SO_4, H_3PO_4)

Method number 1

Application Air analysis

Analytical principle Ion chromatography

Completed in April 1997

Contents

1 General principles

The method described permits the determination of the mists of the inorganic acids phosphoric acid and sulfuric acid, the latter in the range from 0.1 to 2 times the currently valid threshold limit values in air [1, 2].
During sampling, the air containing the acid is drawn through a quartz fibre to collect the acids. Desorption is carried out with an aqueous sodium bicarbonate/sodium hydrogen carbonate solution. Quantitative determination is carried out by ion chromatography.

2 Equipment, chemicals and solutions

2.1 Equipment

Pump for personal sampling, flow rate 210 L/h
Gasmeter
Filter holder, diameter 37 mm, attached to a sampler for total dust
Ion chromatograph with suppressor and conductivity detector
Data processing unit
Quartz fibre filter, diameter 37 mm (e. g. from Ederol)
10 to 1000 mL Volumetric flasks
50 to 10000 µL Adjustable pipettors
10 mL Screw-cap polyethylene vessels
Disposable filter, pore size 0.45 µm
2 to 5 mL Disposable syringes with 60 × 0.6 mm needles
Ultrasonic bath
Ultrapure water system

2.2 Chemicals

Sulfate standard solution β (SO_4^{2-}) = 1.000 g/L (e. g. from Merck, Order No. 19813)
Phosphate standard solution β (PO_4^{3-}) = 1.000 g/L (e. g. from Merck, Order No. 19898)
Sodium bicarbonate anhydrous, analytical grade, (e. g. from Fluka, Order No. 71350)
Sodium hydrogen carbonate, analytical grade, (e. g. from Fluka, Order No. 71627)

2.3 Solutions

The following solutions are prepared using ultrapure water (resistivity >17 MΩ · cm).

Stock solution of the eluents:
2.86 g Na_2CO_3 and 0.25 g $NaHCO_3$ are diluted in water in a 100 mL volumetric flask and the flask is filled to the mark.

Eluent and elution solution:
10 mL of the stock solution is placed in a 1000 mL volumetric flask and the flask is filled to the mark with ultrapure water (0.0027 M Na_2CO_3, 0.0003 M $NaHCO_3$).

2.4 Calibration standards

Combined standard:
4 mL of each of the sulfate and phosphate standard solutions (β = 1 g/L) are placed in a 20 mL volumetric flask and the flask is filled to the mark with ultrapure water. The combined standard solutions thus produced have a concentration of 200 mg/L per anion.

Calibration solutions:
From the combined standard solution, the following calibration solutions are produced by dilution:

Tab. 1. Calibration solutions.

Calibration solution No.	Volume of the combined standard µL	Final volume of the calibration solution mL	Concentration Sulfate mg/mL	Concentration Phosphate mg/mL
1	80	20	0.8	0.8
2	200	20	2.0	2.0
3	400	20	4.0	4.0
4	600	20	6.0	6.0
5	900	20	9.0	9.0
6	1200	20	12.0	12.0
7	1400	20	14.0	14.0
8	1600	20	16.0	16.0
9	1800	20	18.0	18.0
10	2000	20	20.0	20.0

The combined standards can be stored at a temperature of 4 °C for at least one year; the calibration solutions must be freshly prepared when needed.

3 Sample collection and preparation

A filter holder containing a quartz fibre filter is used for sampling. With a pump equipped with a flow regulator, air is drawn through the filter at a flow rate of 210 L/h. The sampling head must fulfil the requirements of DIN EN 481 for inhalable aerosols [4]. Immediately after sampling, the filter placed in a screw-cap vessel, covered with 4 mL desorption solution and closed with the plastic cap that fits. Shake the vessel carfully.

The vessel can be stored in a refrigerator for at least four weeks. For analysis the filter is finally treated for 15 minutes in an ultrasonic bath. The vessels are then left to cool for about 30 minutes. The sample solution is then filtered through a disposable filter and injected into the ion chromatograph.

4 Operating conditions for ion chromatography

Column:	200 × 4.0 mm IonPac AS12A, from Dionex
Precolumn:	50 × 4.0 mm IonPac AS12A, from Dionex
Suppressor:	ASRS (4 mm) with SRS controller, from Dionex
Detector:	Conductivity detector
Solvent:	0.0027 M Na_2CO_3, 0.0003 M $NaHCO_3$ in ultrapure water
Pressure:	approx. 1.1 10^5 hPa
Flow rate:	1.5 mL/min
Injection volume:	50 µL

Figure 1 shows an example of a chromatogram obtained under the conditions given above.

5 Analytical determination

A volume of 50 µL of the sample solution prepared as described in Section 3 is injected into the injection loop of the ion chromatograph with a microliter syringe and analysed under the conditions given above.
If the analytical results are not within the calibration range, the samples must be diluted appropriately and analysed again.

6 Calibration

The calibration solutions described in Section 2.4 are used to draw a calibration curve. Volumes of 50 µL of each of the calibration solutions are injected and analysed as for the sample solutions. The peak areas obtained are plotted against the corresponding concentrations. The calibration curves for sulfuric acid and phosphoric acid are linear in the range from 0.8 mg/L to 20 mg/L. Figure 2 shows examples of the calibration curves.
To check the calibration curve, a control sample should be analysed each day. The calibration curve must be prepared again if the analytical conditions change or quality control shows this to be necessary.

7 Calculation of the analytical result

The concentrations of sulfuric acid and phosphoric acid in the workplace air are calculated using the concentrations of the substances in the solution calculated by the data processing unit. The data processing unit uses the calibration function calculated from the calibration curve.
The concentrations of the acids in the workplace air are calculated from the concentrations of the solutions, taking into account dilution factors and the sampled air volume.
The peak area of the sample signal is calculated according to the following equation:

$$F_{cor} = F - F_{blank}$$

The following equations apply for the concentration of the inorganic acids in the workplace air:

$$\rho = \frac{(F_{cor} - a) \cdot f_{acid} \cdot 0.01\,\text{L}}{b \cdot V_{air}} \cdot \frac{273 + t_g}{273 + t_a}$$

At 20 °C and 1013 hPa:

$$\rho_o = \rho \, \frac{273 + t_a}{293} \cdot \frac{1013}{p_a}$$

where:

F	Peak area from the sample chromatogram
F_{blank}	Peak area for the blank value
F_{cor}	Peak area after correction for the blank value
ρ	Concentration by weight of the inorganic acids in the ambient air as a function of t_a and p_a
ρ_o	Concentration by weight of the inorganic acids in the ambient air at 20 °C and 1013 hPa
a	Intercept of the calibration curve with the ordinate
b	Gradient of the calibration curve in L/mg
f_{acid}	Stoichiometric conversion factor $X^- \rightarrow HX$
0.01 L	Conversion factor for the volume of the measured sample
V_{air}	Sampled air volume in m^3
t_g	Temperature in the gasmeter in °C
t_a	Temperature of the ambient air in °C
p_a	Atmospheric pressure of the ambient air in hPa

Where appropriate when calculating the analytical results, the recovery must be taken into account.

8 Reliability of the method

The characteristics of the method were determined according to the standard DIN EN 482 [5].

8.1 Precision

To calculate the precision of repeated analyses, 10 spiked quartz fibre filters were analysed at each of four concentrations. A microliter syringe was used to spike the quartz fibre filter directly.

Tab. 2. Standard deviation (rel.) *s* and mean variation *u*, *n* = 10 determinations.

Substance	Concentration mg/m³	Standard deviation (rel.) *s* %	Mean variation *u* %
H_2SO_4	0.01	2.6	5.8
	0.02	1.6	3.6
	0.08	0.5	1.1
	0.20	0.6	1.3
	1.00	0.9	2.0
H_3PO_4	0.01	3.2	7.1
	0.08	1.4	3.1
	0.20	0.7	1.6
	1.00	2.4	5.3
	2.00	2.9	6.4

8.2 Recovery rate

The recovery rate was tested for the whole calibration range. A calibration with calibration solution (see Section 2.4) and a calibration with spikes quartz fibre filters were compared. The recovery rate for both acids was a constant 97–100% in the whole calibration range.

8.3 Quantification limit

To determine the quantification limits the range of scatter of the blank value of the quartz fibre filters was evaluated and a calibration curve in the lower concentration range was drawn [6]. The quantification limits for a sampled air volume of 420 L are 0.01 mg/m³ for both substances.

8.4 Sample stability

To determine the sample stability, a series of quartz fibre filters (n = 10) were spiked with defined amounts of sulfuric acid and phosphoric acid. The loaded quartz fibre filters were stored at first for one week at room temperature and then at 4 °C in the refrigerator. Two filters were analysed on the day of loading and after each of 7, 14, 21 and 28 days. No changes in the acid concentrations were detected during this period.

8.5 Specificity

Ion chromatographic determination is highly specific for the anions sulfate and phosphate. Potentially interfering organic acids are separated chromatographically from the anions.

9 Discussion of the method

Both sulfates and phosphates are found ubiquitously in the ambient air. The procedure does not allow differentiation between anions found ubiquitously and those found at the workplace. During sampling care must be taken that the values determined are for the free acids.
The ubiquitous presence of sulfates and phosphates also leads to blank values being produced by numerous materials and equipment (chemicals, glassware). The blank values of all chemicals and equipment must be carefully checked.
Particular thanks goes to the Bundesanstalt für Arbeitschutz und Arbeitsmedizin (BAuA, Dortmund), for their support in finding a suitable filter material.

10 References

[1] *TRGS* 900 „Grenzwerte in der Luft am Arbeitsplatz-Luftgrenzwerte". Bundesarbeitsblatt 10/2000, S. 34–63.
[2] *Deutsche Forschungsgemeinschaft.* List of MAK und BAT Values 2000. Commission for the Investigation of Health Hazards of Chemical Compounds in the Work Area. Report No. 36. WILEY-VCH-Verlagsgesellschaft, Weinheim.
[3] *American Conference of Governmental Industrial Hygienists* (ACGIH): Threshold Limit Values for Chemical Substances and Physical Agents, Cincinnati 1996.
[4] *Europäisches Komitee für Normung* (CEN): DIN EN 481-Arbeitsplatzatmosphäre-Festlegung der Teilchengrößenverteilung zur Messung luftgetragener Partikel. Brüssel 1993. Beuth Verlag, Berlin 1993.
[5] *Europäisches Komitee für Normung* (CEN): DIN EN 482-Arbeitsplatzatmosphäre-Allgemeine Anforderungen an Verfahren zur Messung von chemischen Arbeitsstoffen. Brüssel 1994. Beuth Verlag, Berlin 1994.

[6] *Deutsches Institut für Normung e.V.* (DIN): DIN 32645-Chemische Analytik-Nachweis-, Erfassungs- und Bestimmungsgrenze. Beuth Verlag, Berlin 1994.

Author: *D. Breuer*
Examiners: *R. Hebisch, K. Macho*

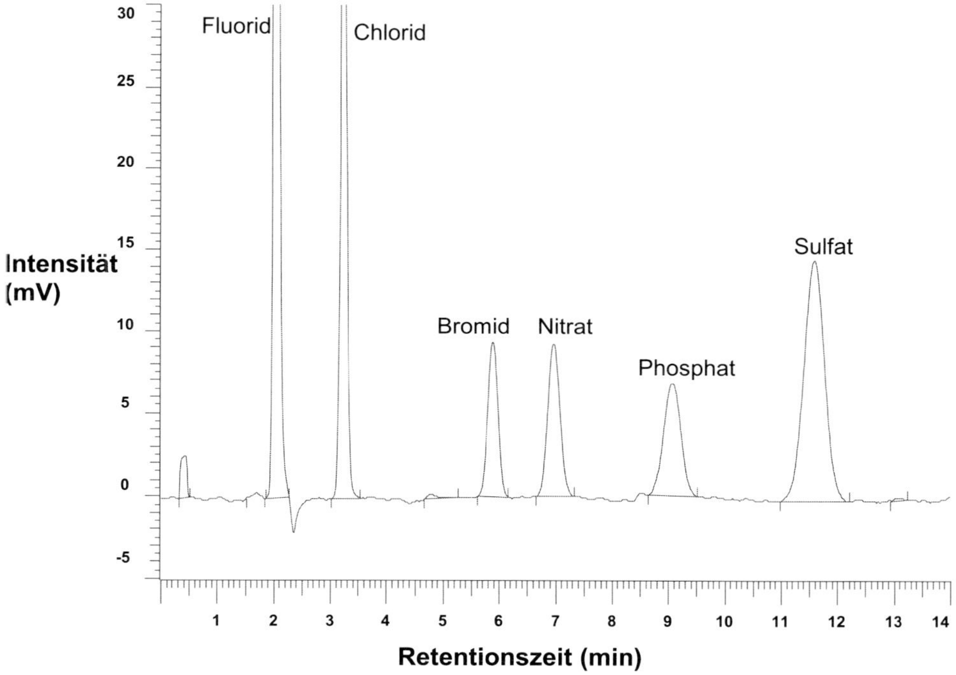

Fig. 1. Example of a chromatogram for the separation of inorganic anions (chromatographic conditions see Section 4). The concentrations are 2.1 mg/L for F^-; Cl^-; Br^-; NO_3^- and 4.2 mg/L for SO_4^{2-}; PO_4^{3-}.

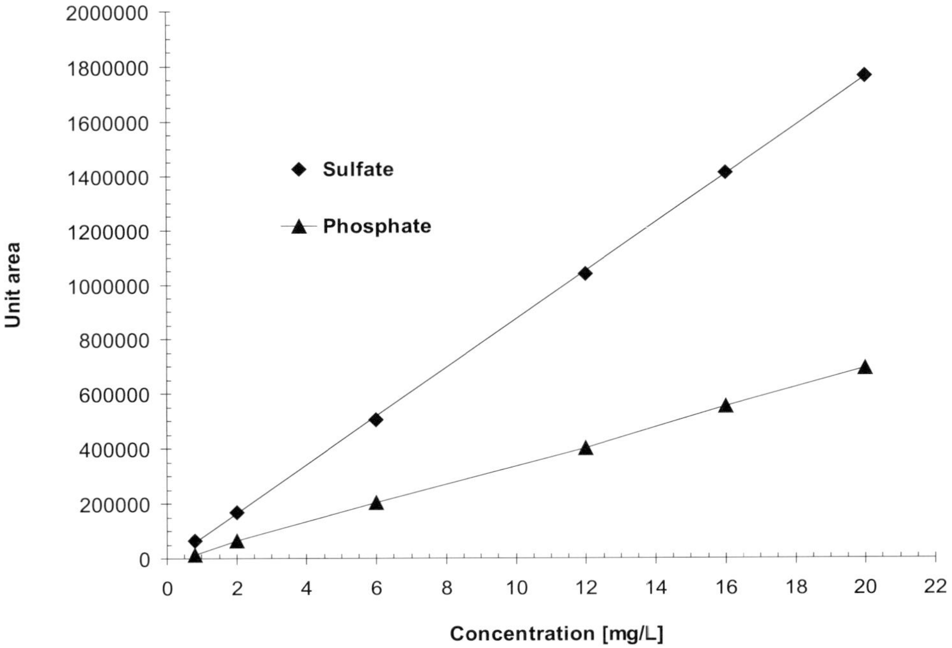

Fig. 2. Calibration curve for inorganic acids (sulfate, phosphate).

Methabenzthiazuron

Method number	1
Application	Air analysis
Analytical principle	High performance liquid chromatography
Completed in	November 1997

Summary

To determine the levels of the herbicide methabenzthiazuron (1-benzothiazol-2-yl-1,3-dimethylurea) in the air, measured air volumes are drawn through Tenax adsorption tubes by a battery-operated pump. The adsorbed substance is extracted with methanol and determined using a UV detector after liquid chromatographic separation. Quantitative evaluation is carried out using a calibration curve. The peak areas obtained using calibration standards are plotted against the methabenzthiazuron concentrations used.

Precision:	Standard deviation (rel.):	s = 2.0 or 3.2%
	Mean variation:	u = 4.6 or 7.4%
	for amounts of methabenzthiazuron of 2.0 and 105 µg per m^3 air and n = 8 determinations	
Limit of quantification:	2 µg/m^3 methabenzthiazuron for a sampled air volume of 720 L	
Recovery rate:	$\bar{\eta}$ = 0.93 (93 %)	
Sampling recommendation:	Sampling time:	6 hours
	Sampled air volume:	720 L

Methabenzthiazuron [CAS No. 18691–97-9]

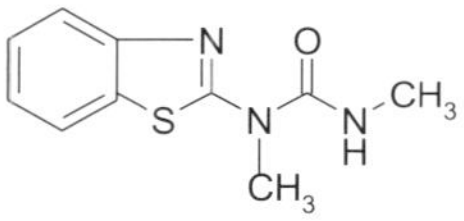

Methabenzthiazuron is a colourless, crystalline solid (molecular weight 221.3, vapour pressure $5.9 \cdot 10^{-8}$ hPa at 20 °C) [1]. Methabenzthiazuron dissolves readily in dichloromethane, toluene and isopropyl alcohol; in water it is sparingly soluble ($5.9 \cdot 10^{-3}$ g/100 g water at 20 °C). The urea derivative methabenzthiazuron is used as a herbicide and is distributed e. g. by BAYER under the trade name Tribunil®.

Author: *K. Riegner*
Examiner: *W. Kleiböhmer*

Methabenzthiazuron

Method number 1

Application Air analysis

Analytical principle High performance liquid chromatography

Completed in November 1997

Contents

1 General principles

To determine the levels of the herbicide methabenzthiazuron in the air measured air volumes are drawn through Tenax® adsorption tubes by a battery-operated pump. The adsorbed substance is extracted with methanol and determined using a UV detector after liquid chromatographic separation. Quantitative evaluation is carried out using a calibration curve. The peak areas obtained using calibration standards are plotted against the methabenzthiazuron concentrations used.

2 Equipment, chemicals and solutions

2.1 Equipment

Adsorption tube type NIOSH, Tenax® (e.g. Günther Karl OHG, Gau-Algesheim, Germany Order No. GK-26–35-03-GO)
Sampling pump (e.g. HI FLOW Sampler, model HFS 113 A, Gilian Instruments, New Jersey, USA)
Gasmeter
Thermometer
Barometer
High performance liquid chromatograph with gradient pump system and UV detector (270 nm)
Glass cutter
50 µL, 100 µL and 250 µL precision syringes
10 mL vials with snap-on caps
Analytical balance
Laboratory shaker
5 mL pipette
5 mL, 10 mL and 25 mL volumetric flasks

2.2 Chemicals

Methabenzthiazuron, w = 99.8% (e.g. Ehrenstorfer, Augsburg, Germany)
Methanol, suitable for HPLC gradient elution
Water, suitable for HPLC
o-Phosphoric acid, w = 85% (e.g. Suprapur, Merck, Darmstadt, Germany)

2.3 Solutions

Stock solution:
To prepare a stock solution, 15 mg methabenzthiazuron is weighed exactly into a 10 mL volumetric flask. The volumetric flask is then filled to the mark with methanol (1.5 g/L).

Elution solution:
To prepare the elution solution 0.5 mL of the stock solution is pipetted into a 25 mL volumetric flask containing approximately 5 mL methanol. The flask is then filled to the mark with methanol (30 mg/L).

Calibration standards:
To check the linearity, calibration standards are prepared from the stock solution or elution solution by transferring the volumes listed in Table 1 to 5 mL volumetric flasks and diluting to the mark with methanol. The following pipetting scheme is used (Table 1):

Table 1. Pipetting scheme for the preparation of the calibration standards.

Concentration of the elution or stock solution mg/L	Volume of the elution or stock solution µL	Final volume of the calibration standards mL	Concentration of the calibration standards mg/L
30	25	5	0.15
30	50	5	0.30
30	100	5	0.60
30	150	5	0.90
30	250	5	1.50
1500	25	5	7.50
1500	50	5	15.00

3 Sample collection and preparation

The adsorption tube is opened at both ends and connected to the inlet of the pump with a short tube. The adsorption tube contains two adsorption phases separated by cotton wool; the larger phase (100 mg Tenax®) is closest to the inlet of the tube during air sampling. The second, smaller adsorption phase (50 mg Tenax®) serves to check for breakthrough of the substance during sampling. For sampling, air is drawn through the adsorption tube at a flow rate of 2 L/min for six hours. After sampling, the air sample volume is noted. In addition, the parameters important for determining the concentration, such as temperature and air pressure at the site of sampling, must be determined. The closed tubes should be stored in a refrigerator until sample preparation.

For preparation, the collection phase and the control phase are each transferred separately to 10 mL vials with snap-on caps; the cotton wool separating the two phases and the upper piece of cotton wool at the tube inlet are analysed together with the collection phase. The glass tube is rinsed with 5 mL methanol. These 5 mL aliquots of methanol are added to the first phase, another 5 mL methanol is added to the second phase and the vials are closed with snap-on caps. The substance is extracted from the adsorption material by shaking the tube on a laboratory shaker for 20 minutes. 50 _L of the methanol solution is then injected without further processing into the high performance liquid chromatograph for separation.

In each analysis series a reagent blank is prepared and analysed in the same way.

4 Operating conditions for high performance liquid chromatography

Precolumn:	Material:	Lichrospher RP-18 (Select B)
	Length:	4 mm
	Internal diameter:	4 mm
	Particle size:	5 µm
Column:	Material:	Lichrospher RP-18 (Select B)
	Length:	25 cm
	Internal diameter:	4 mm
	Particle size:	5 µm
Column temperature:	40 °C	
Solvent:	A = water (adjusted to pH 4 with phosphoric acid)	
	B = methanol	
Gradient:	0 minutes	10% B
	3 minutes	10% B
	4 minutes	70% B
	13 minutes	90% B
	18 minutes	90% B
	19 minutes	10% B
	24 minutes	10% B
Flow rate:	1.0 mL/min	
Detector:	UV detector	
Detection wavelength:	270 nm	
Injection volume:	50 µL	

Under the chromatographic conditions described above, the retention time was about 11.2 minutes (see Figure 1).

5 Analytical determination

Under the conditions described above, 50 µL of the methanol solution is injected into the high performance liquid chromatograph using an injection loop. After high performance liquid chromatographic separation of the methabenzthiazuron from the other components collected on the Tenax and extracted with methanol, the methabenzthiazuron is detected using a UV detector.

6 Calibration

50 µL of each of the calibration standards (see Section 2.3) is injected into the high performance liquid chromatograph and detected using UV detector. To draw the calibration curve the measured peak areas or heights are corrected by subtraction of the reagent blank values and then plotted against the methabenzthiazuron concentrations used in mg/L (see Figure 2). Each solution is analysed twice, and the mean value is used for calculation.
The linearity of the detector has been checked and verified in the injection concentration range from 0.144–15.12 mg/L.

7 Calculation of the analytical result

Using the peak areas or heights obtained after subtraction of the reagent blank values, the methabenzthiazuron concentration in mg/L methanol is read from the calibration curve. If analysis of the control phase reveals a methabenzthiazuron concentration of more than 10% of the total methabenzthiazuron concentration (from the collection phase and the control phase), a breakthrough has taken place. In this case sampling must be repeated under different conditions (e. g. a lower air sample volume).
The concentration by weight ρ (mg of methabenzthiazuron per m^3 of air) in the sample air is calculated according to the following equation:

$$\rho = \frac{a \cdot b}{V_Z \cdot \eta} \qquad (1)$$

At 20 °C and 1013 hPa:

$$\rho_0 = \rho \cdot \frac{273 + t_a}{293} \cdot \frac{1013}{p_a} \qquad (2)$$

The corresponding concentration by volume – independent of the variables pressure and temperature – is given by:

$$\sigma = \rho_0 \cdot \frac{V_m}{M} = \rho \cdot \frac{273 + t_a}{293} \cdot \frac{1013}{p_a} \cdot \frac{24.1}{221.3} \qquad (3)$$

$$\sigma = \rho \cdot \frac{273 + t_a}{p_a} \cdot 0.377$$

For $t_a = 20$ °C and $p_a = 1013$ hPa:

$$\sigma = \rho \cdot 0.109 \frac{\text{mL}}{\text{mg}} \qquad (4)$$

where:

a is the methabenzthiazuron concentration in methanol in mg/L taken from the calibration curve
b is the volume of methanol used for extraction in L
V_z is the air sample volume in m^3
η is the recovery
t_a is the temperature of the ambient air in °C
p_a is the air pressure of the ambient air in hPa
η is the methabenzthiazuron concentration in the ambient air in mg/m^3 at t_a and p_a
ρ_o is the methabenzthiazuron concentration in the ambient air in mg/m^3 at 20 °C and 1013 hPa
M is the molecular weight of methabenzthiazuron
V_m is the molecular volume in L/mole
σ is the methabenzthiazuron concentration in the ambient air in mL/m^3

8 Reliability of the method

8.1 Precision

To determine the precision, the standard deviation for each of 4 assays was evaluated under different climatic conditions (20 °C/40% relative humidity and 30 °C/70% relative humidity). The concentrations of methabenzthiazuron were 2 and 105 µg/m^3. No significant difference was detected between the standard deviations for either of the concentrations under different climatic conditions; the results have therefore been combined as repeated standard deviation determinations (see Table 2).

Table 2. Standard deviation (relative) s and mean variation u, $n = 8$ determinations.

Concentration µg/m^3	Standard deviation (relative) s %	Mean variation u %
0.002	2.0	4.6
0.105	3.2	7.4

8.2 Recovery rate

To determine the recovery, defined amounts of the substance in dissolved form were added to each of 4 adsorption tubes. The solvent was removed by drawing air through

the tube (2 L/min) for about ten minutes. The adsorption tubes were then subjected to defined climatic conditions. After a short equilibration phase (about 20 minutes) conditioned air was drawn through the tubes for a period of 6 hours at a flow rate of 2 L/min. The tubes were then prepared and analysed as described in Sections 3 and 4. The substance was not detected on the second phase of the Tenax adsorption tube.
Table 3 shows the resulting recovery values.

Table 3. Recovery η.

Concentration mg/m³	Temperature °C	Humidity (relative) %	Recovery %
0.002	20	40	92
0.002	30	70	92
0.105	20	40	92
0.105	30	70	95
Mean value			93

8.3 Quantification limit

With a sampled air volume of 720 L, an extraction volume of 5 mL, and an injection volume of 50 µL methabenzthiazuron concentrations down to 2 µg/m³ air can be determined under the given analytical conditions.

9 Discussion of the method

The method described allows the determination of methabenzthiazuron in air in a concentration range of 0.002–0.105 mg/m³. The procedure was tested under various climatic conditions. It was found that the substance could not be desorbed with air either at low or at high concentrations, temperatures or humidities.
Tenax adsorption tubes are suitable for the determination of methabenzthiazuron in the air, both as particles and in gaseous form [2].
To separate methabenzthiazuron from other components an RP-18 LiChrospher 100 (Merck, Darmstadt) can be used instead of the HPLC column described. Care must be taken when determining the recovery that the solution is added to the middle of the tube and that the spiked volume is 10–50 µL. Smaller volumes increase the inaccuracy of the dose, larger volumes lead, however, to increased substance losses during removal of the solvent.
Instruments used: High performance liquid chromatograph Perkin Elmer series 200, variable UV detector and autosampler

10 References

[1] *Industrieverband Agrar e.V.* (Ed.) (1990). Wirkstoffe in Pflanzenschutz- und Schädlingsbekämpfungsmitteln. BLV-Verlagsgesellschaft mbH, München, p. 273

[2] *Riegner K, Schmitz JR* (1994). Erzeugung einer Testatmosphäre und Abscheidung gasförmiger und partikelgebundener Pflanzenschutzmittelrückstände aus Luft auf Tenax®-Sammelröhrchen. Pflanzenschutz-Nachrichten Bayer 47: Pflanzenschutz-Nachrichten Bayer 47: 161–176

Author: *K. Riegner*
Examiner: *W. Kleiböhmer*

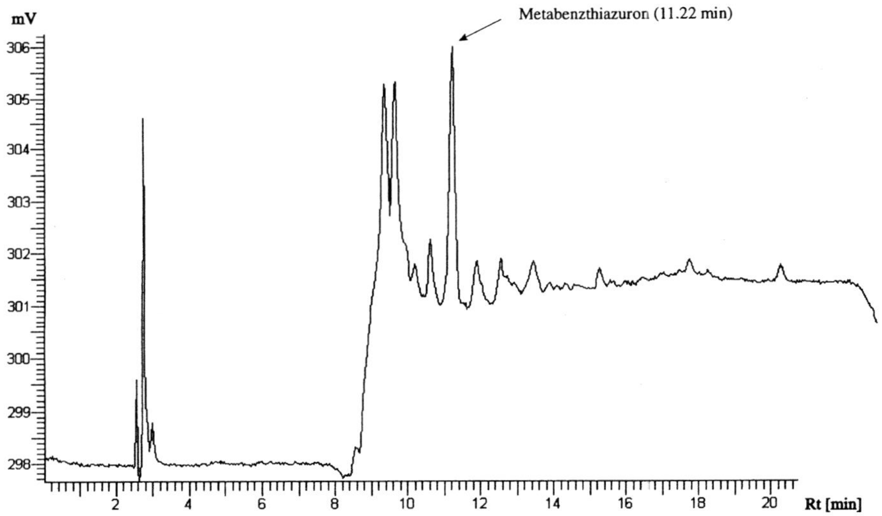

Fig. 1. Example of a calibration standard chromatogram for methabenzthiazuron (14.35 ng absolute). Tenax recovery; 0.002 mg/m3; 20 °C; RH = 40%

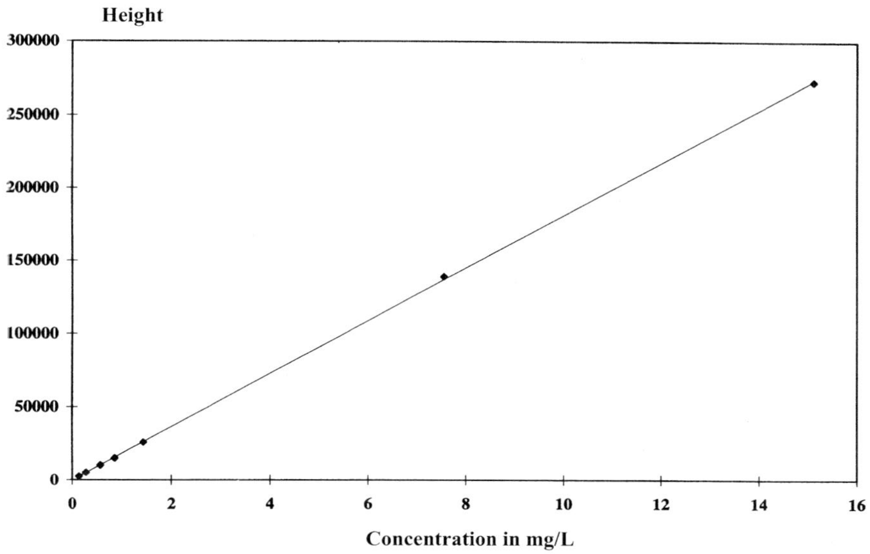

Fig. 2. Example of a calibration curve.

Parathion

Method number	1
Application	Air analysis
Analytical principle	Gas chromatography
Completed in	November 1998

Summary

To determine the levels of the insecticide parathion (O,O–diethyl–O–4–nitrophenyl thiophosphate) in the air, measured air volumes are drawn through Tenax adsorption tubes by a battery-operated pump. The adsorbed substance is extracted with *n*–butyl acetate and determined using a thermionic detector (TSD, NPD) after gas chromatographic separation. Quantitative evaluation is carried out using a calibration curve. The peak areas obtained using calibration standards are plotted against the parathion concentrations used.

Precision:

Table D 1. Standard deviation (relative) s and mean variation u, $n = 6$ determinations.

Concentration µg/m³	Standard deviation (relative) s %	Mean variation u %
0.02	3.2	7.8
4.59	4.5	11.0
174.64	2.0	4.9

Limit of quantification:	0.02 µg/m³ parathion with a sampled air volume of 720 L
Recovery:	$\bar{\eta} = 1.08$ (108%)
Sampling recommendation:	Sampling time: 6 hours Sampled air volume: 720 L

Parathion [CAS No. 56-38-2]

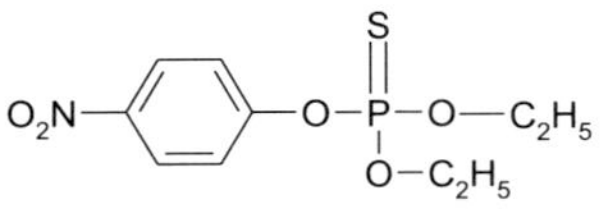

The insecticide parathion (molecular weight 291.3, vapour pressure $8.9 \cdot 10^{-6}$ hPa at 20 °C) is a pale yellow liquid. It has several synonyms (e.g. E 605, thiophos and paraphos). Parathion is used against biting and sucking insects in the growing of crops, vegetables, fruit and grapes. It is readily soluble in alcohol, esters, ethers, ketones and aromatic hydrocarbons; in water, however, it is practically insoluble (20 ppm).
In vivo parathion is a strong inhibitor of specific and unspecific cholinesterases [1]. It is absorbed well via all exposure routes (intestinal tract, lungs, skin, mucous membranes). Oral doses of 1–3 mg/kg can be lethal for an adult, and 0.1 mg/kg for a child. The MAK value for parathion is 0.1 mg/m^3 (2001) measured as the inhalable aerosol fraction [2]. Because it penetrates intact skin unusually readily, parathion is marked with an "H" in the List of MAK and BAT Values [3].

Author: *K. Riegner*
Examiner: *W. Kleiböhmer*

Parathion

Method number 1

Application Air analysis

Analytical principle Gas chromatography

Completed in November 1998

Contents

1 General principles

To determine the levels of the insecticide parathion in the air measured air volumes are drawn through Tenax adsorption tubes by a battery-operated pump. The adsorbed substance is extracted with *n*–butyl acetate and determined using a thermionic detector (NPD, TSD) after gas chromatographic separation. Quantitative evaluation is carried out using a calibration curve. The peak areas obtained using calibration standards are plotted against the parathion concentrations used.

2 Equipment, chemicals and solutions

2.1 Equipment

Adsorption tube type NIOSH, Tenax® (e.g. Günther Karl OHG, Gau-Algesheim, Germany Order No. GK-26–35-03-GO)
Sampling pump (e.g. HI FLOW Sampler, model HFS 113 A, Gilian Instruments, New Jersey, USA)
Gasmeter
Thermometer
Barometer
Gas chromatograph with NPD
Glass cutter
20 µL, 100 µL, 250 µL and 500 µL precision syringes
5 mL vials with snap-on caps
Analytical balance
Ultrasonic bath
1 mL and 5 mL bulb pipettes
10 mL and 100 mL volumetric flasks
Membrane filter attachment for syringe

2.2 Chemicals

Parathion, $w = 99.5\%$ (e.g. Ehrenstorfer, Augsburg, Germany)
n-Butyl acetate, analytical grade (e.g. Merck, Darmstadt, Germany)

2.3 Solutions

Stock solution:
To prepare a stock solution, 10 mg parathion is weighed exactly into a 10 mL volumetric flask. The flask is then filled to the mark with *n*-butyl acetate (1 g/L).

Elution solution:
To prepare the elution solution 0.1 mL of the stock solution is transferred with a pipette to a 10 mL volumetric flask containing about 5 mL *n*-butyl acetate. The flask is then filled to the mark with methanol (10 mg/L).

Calibration standards:
To check the linearity calibration standards are prepared from the elution solution by transferring the volumes listed in Table to 100 mL and 10 mL volumetric flasks and diluting to the mark with *n*–butyl acetate. The following pipetting scheme is used (Table 1):

Table 1. Pipetting scheme for the preparation of the calibration standards.

Volume of the elution solution µL	Final volume of the calibration standards mL	Concentration of the calibration standard µg/mL
20	100	0.002
100	10	0.100
250	10	0.250
500	10	0.500
1000	10	1.000
1500	10	1.500
2000	10	2.000

3 Sample collection and preparation

The adsorption tube is opened at both ends and connected to the inlet of the pump a short tube. The adsorption tube contains two adsorption phases separated by cotton wool; the larger phase (100 mg Tenax) is closest to the inlet of the tube during air sampling. The smaller adsorption phase (50 mg Tenax) serves for to check for possible breakthrough of the substance during sampling. For sampling, air is drawn through the adsorption tube at a flow rate of 2 L/min for six hours. After sampling, the air sample volume is noted. In addition, the parameters important for determining the concentration, such as temperature and air pressure at the site of sampling, must be determined. The closed tubes should be kept in a refrigerator until sample preparation.

For preparation, the collection phase and the control phase are each transferred separately to 5 mL vials with snap-on caps; the cotton wool separating the two phases and the upper piece of cotton wool at the tube inlet are analysed together with the collection phase. The glass tube is rinsed with 3 mL *n*-butyl acetate. These 3 mL aliquots of *n*-butyl acetate are added to the first phase, another 3 mL *n*-butyl acetate is added to the second phase and the vials are closed with snap-on caps. The substance is extracted from the adsorption material by treatment in an ultrasonic bath for 10 minutes. 2 µL of the supernatant n-butyl acetate solution is used for the gas chromatographic separation.

In each analysis series a reagent blank is prepared and analysed in the same way.

4 Operating conditions for gas chromatography

Column:	Material:	Quartz capillary
	Internal diameter:	0.32 mm
	Length:	50 m

	Stationary phase:	HP Ultra 2
	Film thickness:	0.52 µm
Detector:	NPD	
Temperatures:	Furnace:	120 °C; 3 min, 10 °C/min up to 190 °C, 20 °C/min up to 250 °C; 7 min
	Detector:	300 °C
Injector:	on-column	
Carrier gas:	Helium	3.9 mL/min
Injection volume:	2 µL	

Under the gas chromatographic conditions described above, the retention time was about 14.7 min (see Figure 1).

5 Analytical determination

Under the operating conditions described above, 2 µL of the *n*–butyl acetate solution is injected into the gas chromatograph. To protect the injector system and the GC column it is recommended that any Tenax residue in the *n*–butyl acetate solution be removed before injection using a membrane filter attached to a syringe. If the analytical results are above the calibration range, the solution must be diluted and analysed again.

6 Calibration

2 µL of each of the calibration standards (see Section 2.3) are injected into the gas chromatograph and determined using an NPD detector. To prepare the calibration curve the measured peak areas or heights are plotted against the parathion concentration in mg/L after subtraction of the peak areas or heights measured for the reagent blank (see Figure 2).
The linearity of the detector has been checked and verified in the injection concentration range from 0.002–2 mg/L.

7 Calculation of the analytical result

Using the peak areas or heights obtained after subtraction of the reagent blank values, the parathion concentration in mg/L *n*–butyl acetate is read from the calibration curve. If analysis of the control phase reveals a parathion concentration of more than 10% of the total parathion concentration (from the collection and control phases), breakthrough

has taken place. In this case sampling must be repeated under different conditions (e.g. smaller air sample volumes).
The concentration by weight ρ (mg of parathion per m^3 of air) in the sample air is calculated according to the following equation:

$$\rho = \frac{a \cdot b}{V_Z \cdot \eta} \tag{1}$$

At 20 °C and 1013 hPa:

$$\rho_0 = \rho \cdot \frac{273 + t_a}{293} \cdot \frac{1013}{p_a} \tag{2}$$

The corresponding concentration by volume – independent of the variables pressure and temperature – is given by:

$$\sigma = \rho_0 \cdot \frac{V_m}{M} = \rho \cdot \frac{273 + t_a}{293} \cdot \frac{1013}{p_a} \cdot \frac{24.1}{291.3} \tag{3}$$

$$\sigma = \rho \cdot \frac{273 + t_a}{p_a} \cdot 0.286$$

For $t_a = 20$ °C and $p_a = 1013$ hPa:

$$\sigma = \rho \cdot 0.083 \, \frac{\text{mL}}{\text{mg}} \tag{4}$$

where:

- a is the parathion concentration in *n*-butyl acetate in mg/L taken from the calibration curve
- b is the volume of *n*-butyl acetate used for extraction in L
- V_z is the air sample volume in m^3
- η is the recovery
- t_a is the temperature of the ambient air in °C
- p_a is the air pressure of the ambient air in hPa
- ρ is the parathion concentration in the ambient air in mg/m^3 at t_a and p_a
- ρ_o is the parathion concentration in the ambient air in mg/m^3 at 20 °C and 1013 hPa
- M is the molecular weight of parathion
- V_m is the molecular volume in L/mole
- σ is the parathion concentration in the ambient air in mL/m^3

8 Reliability of the method

8.1 Precision

The precision was calculated by determining the recovery. From 6 assays in each case, in which the amount of substance used was between 0.02 and 131 µg, the standard deviations (relative) *s* and mean variations *u* listed in Table 2 were found.

Table 2. Standard deviation (rel.) *s* and mean variation *u*, *n* = 6 determinations.

Concentration µg/m³	Standard deviation (relative) *s* %	Mean variation *u* %
0.02	3.2	7.8
4.59	4.5	11.0
174.64	2.0	4.9

8.2 Recovery rate

To determine the recovery, defined amounts of the substance in dissolved form were added to each of 6 adsorption tubes. The solvent was removed by drawing air through the tube (2 L/min) for about ten minutes. To determine the recovery, air was drawn through the tube for a period of 6 hours at a flow rate of 2 L/min. Then the samples were prepared and analysed as described in Sections 3 and 4.
Table 3 shows the resulting recovery values.

Table 3. Recovery η

Concentration µg/m³	Recovery η %
0.02	115
4.59	92
174.64*	117
Mean value	108

* The amounts of the substance detected on the control phase of the Tenax adsorption tube were less than 0.1% of the spiked amounts.

8.3 Quantification limit

With a sampled air volume of 720 L, an extraction volume of 5 mL and an injection volume of 2 µL parathion concentrations down to 0.02 $\mu g/m^3$ air can be determined under the given analytical conditions.

8.4 Shelf-life

To check the shelf-life, each of 3 Tenax tubes were spiked with defined amounts of parathion. For a sample volume of 720 L, the spiked amounts corresponded to air concentrations of about 0.02 $\mu g/m^3$ and 5 $\mu g/m^3$. The solvent was removed by drawing air through the tube (2 L/min) for about ten minutes. The loaded tubes were stored for 12 days at +5 °C. The samples were then prepared and analysed as described in Sections 3 and 4. During this period no changes in the parathion concentration could be detected.

9 Discussion of the method

The method described allows the determination of parathion in air in a concentration range of 0.02–175 $\mu g/m^3$. Tenax adsorption tubes are suitable for the determination of parathion in the air, both as particles and in gaseous form [4].
As an alternative to thermionic detection, selected ion monitoring can be used. The analytical characteristics must then be checked.
Care must be taken when determining the recovery that the solution is added to the middle of the tube and that the spiked volume is 10–50 µL. Smaller volumes increase the inaccuracy of the dose, larger volumes lead, however, to increased substance losses during removal of the solvent.
To separate parathion from other components, in addition to the described GC column, the HP Ultra 1 can be used. As an alternative to on-column loading, injection of the extract at 260 °C is possible.
Changes in the detector sensitivity during an analytical sequence can be compensated for by the addition of an internal standard (e. g. 4-nitrobenzyl bromide) before gas chromatographic determination.

Instruments used: Gas chromatograph HP 6890, NPD with ion source (Blos Analysentechnik, Munich, Germany), Gerstel Kaltaufgabesystem KAS 4 with on-column inlet (Gerstel, Germany)

10 References

[1] *Henschler D* (Ed) (1973) Parathion. In: Gesundheitsschädliche Arbeitsstoffe. Toxikologisch-arbeitsmedizinische Begründungen von MAK-Werten. VCH-Verlagsgesellschaft, Weinheim

[2] *Deutsche Forschungsgemeinschaft* (2001) MAK- und BAT-Werte-Liste 2001. Senatskommission zur Prüfung gesundheitsschädlicher Arbeitsstoffe, Mitteilung 37. WILEY-VCH-Verlag, Weinheim

[3] *Heimann KG* (1994) Ethyl Parathion: Begründung der Bestimmungsgrenze in der Luft. Unpublished report, BAYER AG, Leverkusen

[4] *Riegner K, Schmitz JR* (1994) Erzeugung einer Testatmosphäre und Abscheidung gasförmiger und partikelgebundener Pflanzenschutzmittelrückstände aus Luft auf Tenax®-Sammelröhrchen. Pflanzenschutz-Nachrichten Bayer 47: 161–176

Author: *K. Riegner*
Examiner: *W. Kleiböhmer*

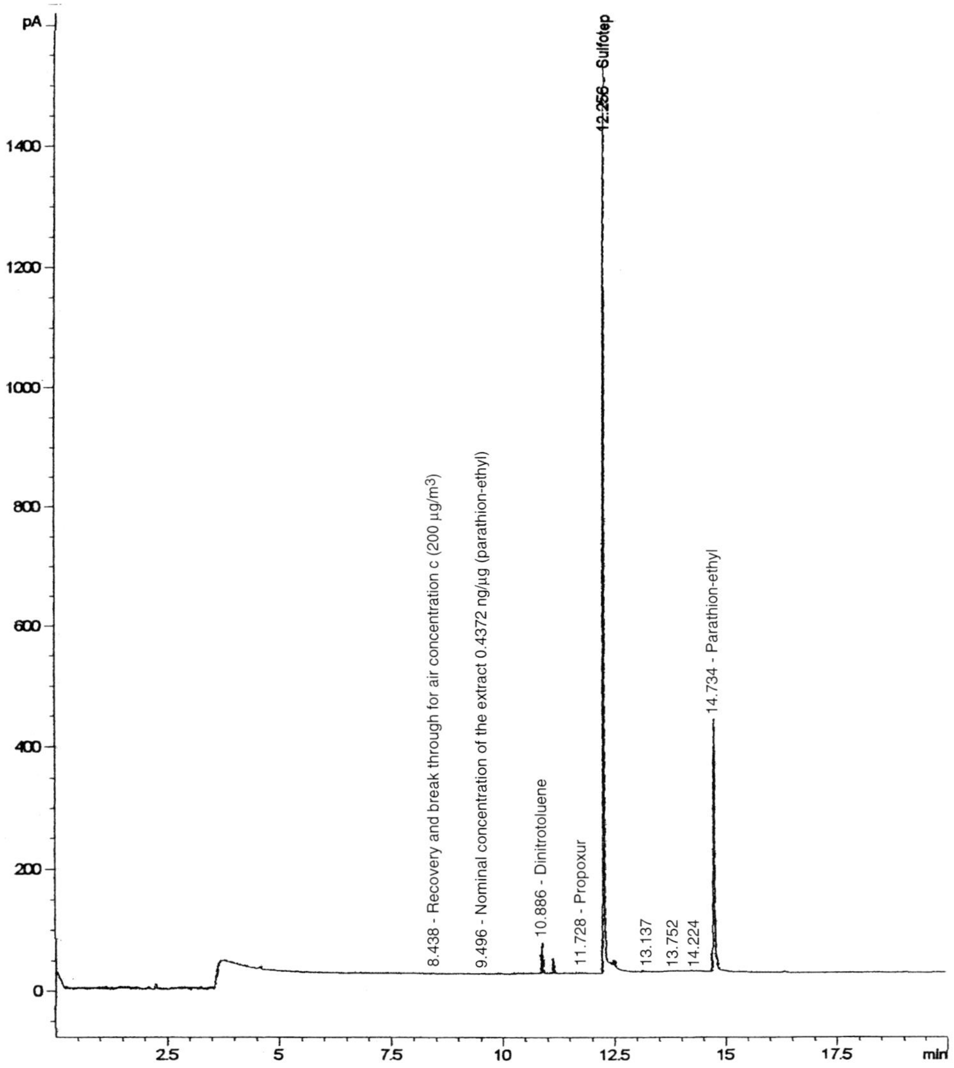

Fig. 1. Example chromatogram for parathion (0.9 ng absolute).

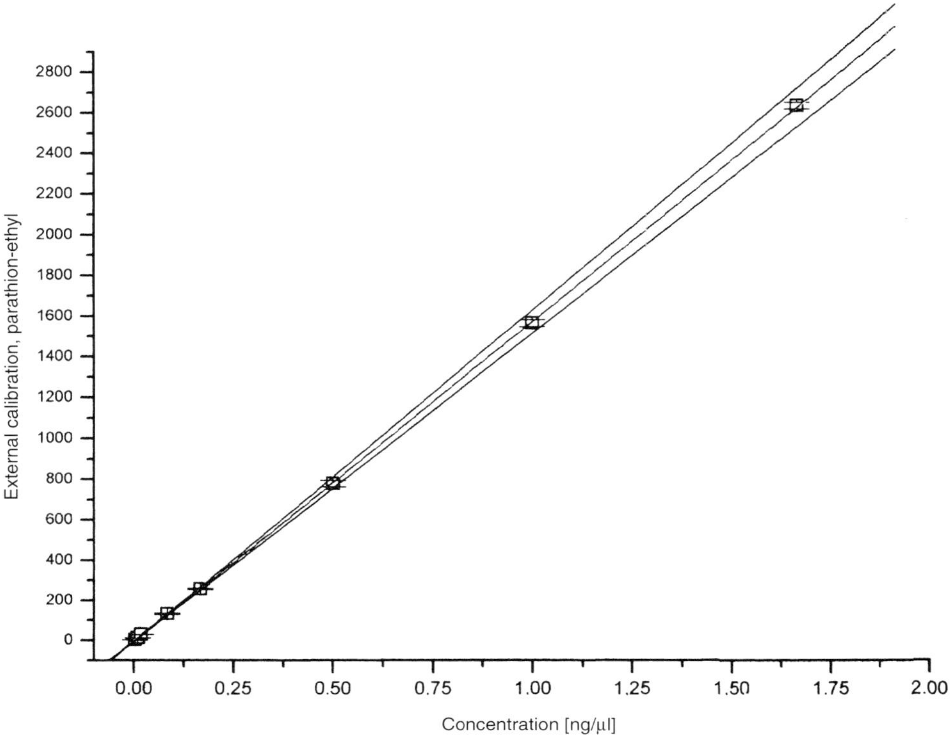

Fig. 2. Example of a calibration curve.

Polyisocyanates made from aromatic diisocyanates

Method number 1

Application Air analysis

Analytical principle High performance liquid chromatography

Completed in November 1995

Summary

This method, unlike method No. 1 Polyisocyanates [1], deals only with polyisocyanates made from aromatic diisocyanates or from mixtures including aromatic diisocyanates. The polyisocyanate-containing substances present at the workplace as an aerosol are drawn with a sampling pump into an adsorption tube containing glass wool impregnated with a secondary amine as derivatisation reagent. During sampling, the droplet aerosol is collected quantitatively on the glass wool. The rate of suction of the air sample into the adsorption tube is set to capture inhalable aerosol particles. The free isocyanate groups of the polyisocyanates react immediately on contact with the reagent to form soluble urea derivatives. Immediately after sampling, these are eluted together with the deposited aerosol in a countercurrent of dichloromethane. During this process any remaining free isocyanate groups also react rapidly to form urea derivatives in the homogeneous phase. The urea derivatives are analysed in a high performance liquid chromatograph equipped with a UV detector. Quantitative evaluation is carried out using a calibration curve. Calibration is carried out with the polyisocyanate type used for the work process being monitored. Polyisocyanates are contained in amounts of about 40–80% w/w in so-called hardeners in two-component systems, and in amounts of >70% w/w in one-component systems. Reaction with the derivatisation reagent is carried out in the same way as described elsewhere [1, 2].

Concentration range:	0.1–10 mg/m^3 polyisocyanate in air
Precision of the whole procedure:	Standard deviation (rel.): s = <10% Mean variation: u = <30% for a weight of ~100 µg polyisocyanate and n = 6 determinations

Limit of quantification:	0.1 mg/m^3 polyisocyanate in air (corresponding to about 3 µg/mL sample solution with a sampled air volume of 30 L, 1 mL of sample solution and a 10 µL injection volume)
Recovery:	$\eta > 0.9$ (> 90%),
Sampling recommendation:	Sampling time: 15 minutes Sampled air volume: 30 L

Polyisocyanates

General notes on this substance group can be found in the corresponding section in Method No. 1, Polyisocyanates, vol. 3 [1]. Products made from aromatic polyisocyanates or mixtures containing aromatic polyisocyanate are used in practice to a lesser extent than the products made from purely aliphatic polyisocyanates described in Method 1 [3]. The aromatic isocyanate used as starting material is toluene diisocyanate (TDI, used in some cases in the form of a technical grade mixture of isomers, 2,4-TDI and 2,6-TDI). The present method is intended only for these polyisocyanates. Depending on the type of polymer, they have different properties (see above) and contain different levels of solvents. The sampling procedure described is suitable for the polyisocyanates listed below in a lacquer formulation with polyalcohols (or alone) under the conditions tested [4]. Control samples serve to check suitability under the conditions found in practice. If it is suspected that the analytical results have been falsified by the too rapid formation of polyurethane, sampling can be checked using a more complicated procedure (e.g. Midget impinger with nitro reagent followed by a filter to collect the aerosol fractions which were not deposited).

Trade names common at present for polyisocyanates made from aromatic diisocyanates are:

Desmodur®	(Bayer)	Tolonate®	(Rhone-Poulenc, France)
Basonat®	(BASF)	Mondur™	(Bayer Co, U.S.A.)

Polyisocyanate made from TDI and hexamethylene diisocyanate (HDI) and having an isocyanurate structure (ideal structure): Desmodur HL.

Desmodur HL

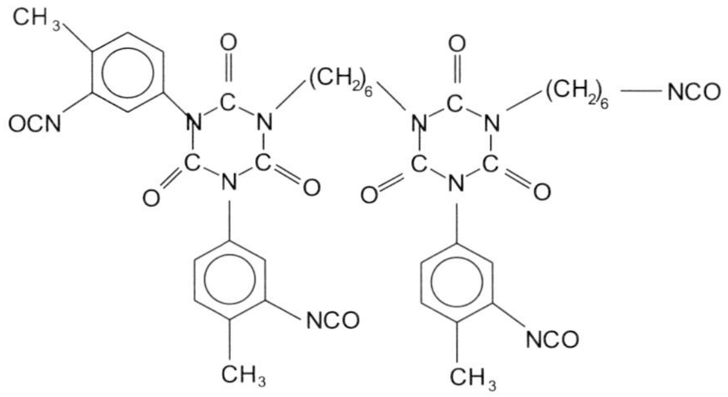

Polyisocyanate made from TDI and having an isocyanurate structure (ideal structure): Desmodur IL.

Desmodur IL

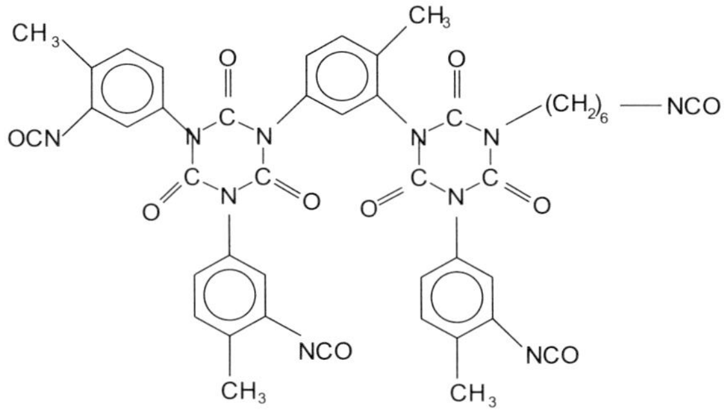

Polyisocyanate made from TDI and having a urethane structure (ideal structure): Desmodur L.

Desmodur L

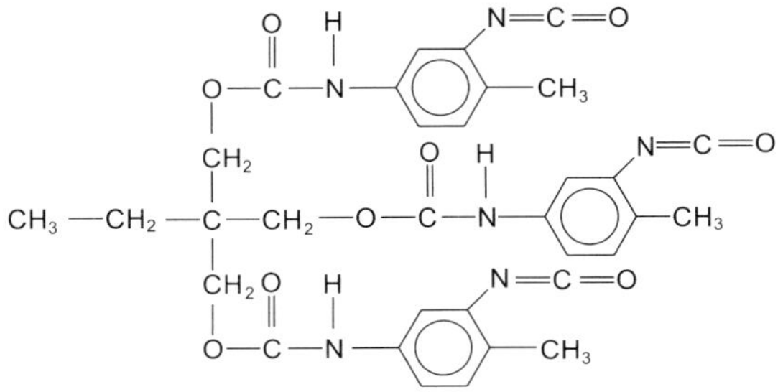

Author: *M. Kuck*
Examiner: *J.U. Hahn*

Polyisocyanates made from aromatic diisocyanates

Method number	1
Application	Air analysis
Analytical principle	High performance liquid chromatography
Completed in	November 1995

Contents

1 General principles

The polyisocyanate-containing substances present at the workplace as an aerosol are drawn with a sampling pump into an adsorption tube containing glass wool impregnated with a secondary amine as derivatisation reagent. During sampling, the droplet aerosol is collected quantitatively on the glass wool. The rate of suction of the air sample into the adsorption tube is set so that aerosol particles corresponding to the definition of inhalable dusts according to DIN EN 481 are sampled [5]. The free isocyanate groups of the polyisocyanates react immediately on contact with the reagent to form soluble urea derivatives. Immediately after sampling, these are eluted together with the deposited aerosol in a countercurrent of dichloromethane. During this process any remaining free isocyanate groups also react rapidly to form urea derivatives in the homogeneous phase. They are analysed in a high performance liquid chromatograph equipped with a UV detector. Quantitative evaluation is carried out using a calibration curve or single-point calibration. Calibration is carried out with the polyisocyanate type used for the work process being monitored. Polyisocyanates are contained in amounts of about 40–80% w/w in so-called hardeners in two-component systems, and in amounts of >70% w/w in one-component systems. Depending how calibration and evaluation is carried out, the analytical result can be given as the polyisocyanate or total isocyanate level in the workplace air. Reaction with the derivatisation reagent is carried out in the same way as in Method No. 1, Polyisocyanates, vol. 3 [1] and Method No. 2 HDI, TDI, vol. 1 [2].

2 Equipment, chemicals and solutions

2.1 Equipment

Liquid chromatograph equipped with UV-VIS detector
Various volumetric flasks
500 mL Separating funnel
Various pipettes
Apparatus for evaporation in a stream of nitrogen or a vacuum
Flow-stabilised sampling pump, output at least 2 L/min
(e.g. GSA 2000 from the Gesellschaft für Staubmeßtechnik und Arbeitsschutz mbH, Neuss)
Calibration apparatus for setting the output of the pump, suitable for flow rates of 2 L/min

Thermometer
Barometer
Glass tubing (8 mm external diameter) for producing adsorption tubes, length 70–80 mm, with an internal diameter of about 6 mm, a glass wool filling of 0.5 g long-fibre purified glass wool, providing an adsorption zone 50 mm in length.
Preparation vials for storing samples (e. g. 40 mL EPA vials, Order No. 279525 from ASS-CHEM GmbH, Bad Homburg)

2.2 Chemicals

Methanol, analytical grade
Dichloromethane, analytical grade
Double distilled water
0.5 N HCl, analytical grade
4-Nitrobenzyl propyl ammonium chloride (nitro reagent)
(e. g. Riedel-de-Haen No. 33487 or DFG Method No. 2, HDI, TDI)
1 N sodium hydroxide solution, analytical grade
Purified glass wool

2.3 Solutions

2.3.1 Nitro reagent stock solution

0.8 g nitro reagent hydrochloride (corresponds to 0.67 g free base) is dissolved in 50 mL distilled water and extracted three times in a 100 mL separating funnel, each time with 10 mL dichloromethane. The organic (lower) phase is separated from the aqueous phase each time and discarded. 25 mL 1 N sodium hydroxide solution is then added to the aqueous phase. A white precipitate is formed (free base). This aqueous suspension is transferred to a 500 mL separating funnel and extracted twice with 100 mL dichloromethane. The organic phase is separated from the aqueous phase, dried over sodium sulfate and transferred to a 250 mL volumetric flask; the aqueous phase is discarded. The flask is filled to the mark with dichloromethane and an aluminium paper sleeve is placed round the flask to protect the solution from light. This stock solution contains 2.7 mg nitro reagent/mL and must be stored in the refrigerator. It should be freshly prepared after 4 weeks.

2.4 Calibration standards

2.4.1 Calibration standard for qualitative analysis

5 mL of the sample material obtained as described in Section 3.1 is freed of volatile components under reduced pressure (water-jet vacuum pump) for one hour at a maximum temperature of 50 °C in the absence of moisture under inert gas. One drop of the remaining material is reacted as described in Section 2.4.2 with nitro reagent stock solution.

2.4.2 Calibration standard for quantitative analysis

With one-component systems e. g. three drops of the material are pipetted into a 50 mL volumetric flask and weighed, with two-component systems e. g. five drops of the hardener components corresponding to about 100 mg polyisocyanate are used (see also Section 3.1). The volumetric flask is then filled to the mark with nitro reagent stock solution (135 mg nitro reagent/50 mL). Calibration standards containing about 50–500 μg polyisocyanate/10 mL are produced from this solution by diluting with dichloromethane. When unknown hardener material is used as calibration standard, the amount of polyisocyanate it contains must be determined separately, for example by determining the non-volatile residue or titration of the free NCO groups using dibutylamine (according to DIN EN ISO 11909 [6]).

2.5 Adsorption tubes

2.5.1 Preparation of the tubes

From commercially available glass tubing of 8 mm external diameter, pieces of 8 cm in length are cut and one end of each piece is melted with a Bunsen burner to produce an opening with a cross-section of 6 mm. The tubes are filled with 500 mg long-fibre glass wool and the filling is compressed to a total length of about 5 cm so that a dense cylindrical layer of fibres covers the glass walls. Between the narrowed opening and the layer of glass wool there should be a distance of only about 5 mm.

2.5.2 Impregnation of the tubes

With a pipette 1 mL of nitro reagent is transferred to each tube and the tube is tilted and rotated until the layer of glass wool and the walls are completely wetted with the solution. Then the solvent is removed from the tube using a slight vacuum at room temperature and the evacuated system is aerated with inert gas. The tubes are transferred to preparation vials rinsed with inert gas (see Section 2.1) and the vials are sealed (gastight) and stored in the dark. They can be kept in this way for at least 4 weeks.

3 Sample collection and preparation

3.1 Standard sample material

A preparation vial is filled to the brim with the hardener used for the work process (two-component system) or with the ready-made reaction system (one-component system) and sealed with a gastight seal. This sample serves as reference material for the later determination of the level of polyisocyanate or isocyanate groups, and for producing the calibration standards (see Section 2.4) and cannot be stored indefinitely. It should not be stored for more than a week.

3.2 Control sample before sampling

This sample is only needed if the polyisocyanate reaction system has a pot life (usable life) of less than one hour. Immediately before beginning the air sampling, 10 mL of the reagent solution is placed in a preparation vial, 1 drop of the ready-made reaction mixture is added. The vial is then sealed with a gastight seal and the solution homogenised by shaking.

3.3 Sample collection

Sampling is carried out with adsorption tubes. These contain a layer of long-fibre glass wool loaded with nitro reagent. The aerosol components are deposited, adsorbed and in some cases chemically absorbed in the layer.
For sampling, a tube is taken from the preparation vial and connected at the wide opening to the sampling pump, which is set at a flow rate of 2 L/min. At a flow rate of 2 L/min, the 6 mm opening produces a linear suction rate of at least 1.25 m/s for the droplet aerosol and therefore fulfils the criterion for sampling inhalable dusts.
Immediately after sampling, which should be carried out for at least 15 minutes to reach the determined quantification limit, but no longer than 30 minutes to be sure that further reaction of the polyisocyanates to be determined does not falsify the results, the loaded tube is placed back in the preparation vial, inlet first. With a disposable pipette 5 mL methylene chloride is added through the free tube opening and the vial is sealed with a gastight seal. The volume of reagent added is sufficient to elute the glass wool layer in a countercurrent direction and completes the transformation of still free isocyanate groups of the polyisocyanates in the homogeneous phase.

3.4 Control sample after sampling

Immediately after the end of air sampling, 10 mL reagent solution is placed in another preparation vial and 1 drop of the remaining reaction mixture used for the work process is added and homogenised by shaking. The vial is sealed with a gastight seal.

4 Operating conditions for high performance liquid chromatography

Apparatus:	HP 1090 Liquid Chromatograph
Column:	Macherey-Nagel ET 250/8/4 NUCLEOSIL 120-5 C8

Solvent:	A = methanol
	B = water
Gradient:	From 40% B to 0% B in 25 minutes
Flow rate:	1.5 mL/min
UV detector:	272 nm
Injection volume:	10 µL

5 Analytical determination

In the laboratory the sample solution produced as described in Section 3.3 is reduced to 2 mL by evaporation, 2 mL 0.5 N HCl is added and the solution is shaken vigorously for 15 minutes. This binds the excess nitro reagent as the hydrochloride in the aqueous phase. With this procedure the use of buffer solutions in the eluent can be avoided. To prepare the calibration curve, 10 µL of the appropriate dilution of the standard solution (see Section 2.4.2) is injected three times consecutively for each concentration and the mean of the corrected peak areas is used for the calibration point. In alternation with the analytical samples, calibration solutions of comparable concentration are injected for calibration. Evaluation is carried out automatically in modern systems using a peak area integrator. Care must be taken that the concentration of the calibration solution corresponds with the control value (threshold limit value). With an assumed threshold limit value of e.g. 1 mg/m^3 polyisocyanate in air, the corresponding standard solution contains 30 mg/L of polyisocyanate.

6 Calibration

6.1 Determination of the peak profile of the polyisocyanates

The standard solutions produced as described in Sections 2.4.1 and 2.4.2 are analysed directly and the peak patterns are compared visually. The peaks that can be assigned to the polyisocyanates are contained in the qualitative standard. The corresponding peaks of the quantitative standard which are eluted without interfering peaks can be used for quantitative analysis. Which peaks are finally included in the evaluation depends on the concentration of the polyisocyanates in the sample solution. If the concentration is high above the quantification limit, all interference-free peaks whose relative intensities do not differ by more than one power of ten can be used for evaluation. If the concentration is close to the quantification limit, one may be limited to using the most intensive peak. It must be borne in mind that the certainty with which the polyisocyanates may be identified analytically is proportional to the number of peaks in the characteristic peak pattern. If the number of peaks considered is reduced to one, then only one maximum concentration value (worst case), derived from the moderately meaningful retention time, but not the identification can be given as the analytical result.

6.2 Control samples

The chromatograms of the control samples (before and after sampling) are analysed qualitatively by comparing the peak patterns to check visually whether they differ significantly. If there are at least three intensive peaks, a mean value and the corresponding relative standard deviation can be calculated by determining the peak area ratios of the corresponding peaks from the two samples. If the relative standard deviation is above 20%, it can be assumed that the reaction mixture has changed significantly during the processing period.

6.3 Air samples

With the peak areas determined as described in Section 5 the analytical result can be calculated using the external standard method and single-point calibration, or using a calibration curve.

7 Calculation of the analytical result

With the peak areas determined as described in Section 5 the analytical result can be calculated using the external standard method and single-point calibration, and taking into account the amount of polyisocyanate in the calibration standard material as follows:

$$X = \frac{SF_{\mathrm{Air}} \cdot G_{\mathrm{Ref}}}{SF_{\mathrm{Ref}}}$$

where:

SF_{Air} sum of the peak areas of the most intensive peaks which can be clearly assigned to the polyisocyanate from the air sample.

SF_{Ref} sum of the peak areas in the calibration standard for the components used for SF_{Air}.

G_{Ref} weight of polyisocyanate in µg/L calibration standard solution corresponding with SF_{Ref}.

X weight of polyisocyanate in the sample in µg

The corrected concentrations by weight of polyisocyanate are calculated according to the following equation:

$$\rho = \frac{X}{V_{\mathrm{Z}} \cdot \eta} \cdot \frac{273 + t_{\mathrm{g}}}{273 + t_{\mathrm{a}}}$$

$$\rho_o = \rho \frac{(273 + t_a)}{293} \cdot \frac{1013}{p_a}$$

where:

ρ polyisocyanate concentration in the ambient air in mg/m^3 at t_a and p_a
ρ_o polyisocyanate concentration in the ambient air in mg/m^3 at 20 °C and 1013 hPa
V_z volume of the air sample in litres
η recovery
t_g temperature in the gasmeter in °C
t_a temperature of the air sample in °C
p_a air pressure during sampling in hPa

8 Reliability of the method

8.1 Precision

In accordance with the sampling procedure described, an approx. 5% solution (5 g/100 mL) in methylene chloride was prepared from freshly produced reaction mixture from each of the tested two-component systems (Desmodur HL, IL and L) with about 10% polyisocyanate, and 20 µL of this solution, containing about 100 µg polyisocyanate, was loaded onto the front zone of each of six adsorption tubes. Then 30 litres of laboratory air was drawn through each tube at a flow rate of 2 L/min. The tubes were each processed and analysed as described above.

Table 1. Precision of each of $n = 6$ independent analytical values.

Substance	Standard deviation (relative) s %	Mean variation u %
Desmodur L 67	7.10	17.37
Desmodur IL	7.68	18.79
Desmodur HL	7.61	18.62

8.2 Recovery rate

The recovery was determined by calculating the ratios of the loaded weights of polyisocyanate to the analytically determined weights in the sample tubes as described in Section 8.1, under various conditions. The results obtained at an ambient temperature of about 25 °C and a relative humidity of about 65% are shown in Table 2.

Table 2. Recovery calculated from Table 1.

Substance	Recovery %
Desmodur L 67	92.34
Desmodur IL	97.33
Desmodur HL	91.20

The experiment was also carried out with tubes which were not impregnated. The recovery was only 53% although conditions were otherwise the same.
In addition, loaded tubes were produced as described in Section 8.1 and stored for differing periods above the normal period of 30 minutes after sampling (30 L air). These tests were carried out only with Desmodur L, representative for the other test substances.

Table 3. Recovery of Desmodur L as a function of the storage time.

Storage time min	Recovery %
5	101
30	103
60	120
120	104
479	97

8.3 Quantification limit

Under the given analytical conditions the quantification limit is about 0.1 mg/m^3 polyisocyanate for an air sample volume of 30 litres.

9 Discussion of the method

The method is not unrestrictedly suitable for rapid reaction systems for which considerable reaction of the isocyanate groups to form insoluble polymer structures during sampling is to be expected. For reasons of handling, lacquer systems on the above-mentioned basis generally have a sufficiently long pot life (about 1 hour) whereas the sampling time required in practice is about 15–30 minutes. Even after the pot life has been exceeded by more than 100%, these lacquer systems are still completely soluble. The degree of reactivity of the isocyanates is below 10 % and therefore within the range of error of the total procedure. The suitability of the procedure can be checked by taking samples of the reaction mixture before and after the analysis and derivatising them directly. In the subsequent analysis the two samples should be found to contain, within the tolerance range, the same amounts of polyisocyanate. If that is not the case, sampling can perhaps be checked using a more complicated procedure (e. g. Midget impinger with nitro reagent followed by a filter to collect the aerosol fractions which were not deposited).

10 References

[1] Greim H (Ed.) Analytische Methoden zur Prüfung gesundheitsschädlicher Arbeitsstoffe, Luftanalysen. Band 1, Polyisocyanate auf Basis aliphatischer Diisocyanate Meth.-Nr. 1. VCH-Verlagsgesellschaft, Weinheim 1993
[2] Greim H (Ed.) Analytische Methoden zur Prüfung gesundheitsschädlicher Arbeitsstoffe, Luftanalysen. Band 1, HDI, TDI Meth.-Nr. 2. VCH-Verlagsgesellschaft, Weinheim 1988
[3] Bayer AG: Anwendungstechnische Information: Order No.: KL 44270, issued 1.12.1975
[4] Technischer Arbeitskreis der Fachabteilung Autoreparaturlacke im Verband der Lackindustrie e.V.: Konzentrationsmessungen in der Spritzkabinenluft-Concentration Measurements in Spray Booth Atmospheres. Farbe+Lack 11, 911–914 (1987)
[5] Europäisches Komitee für Normung (CEN): DIN EN 481-Arbeitsplatzatmosphäre-Festlegung der Teilchengrößenverteilung zur Messung luftgetragener Partikel. Brüssel 1993. Beuth Verlag, Berlin 1993
[6] Deutsches Institut für Normung e.V. (DIN): DIN EN ISO 11909-Bindemittel für Beschichtungsstoffe, allgemeine Prüfverfahren, Polyisocyanate. Beuth Verlag, Berlin 1998

Author: *M. Kuck*
Examiner: *J.U. Hahn*

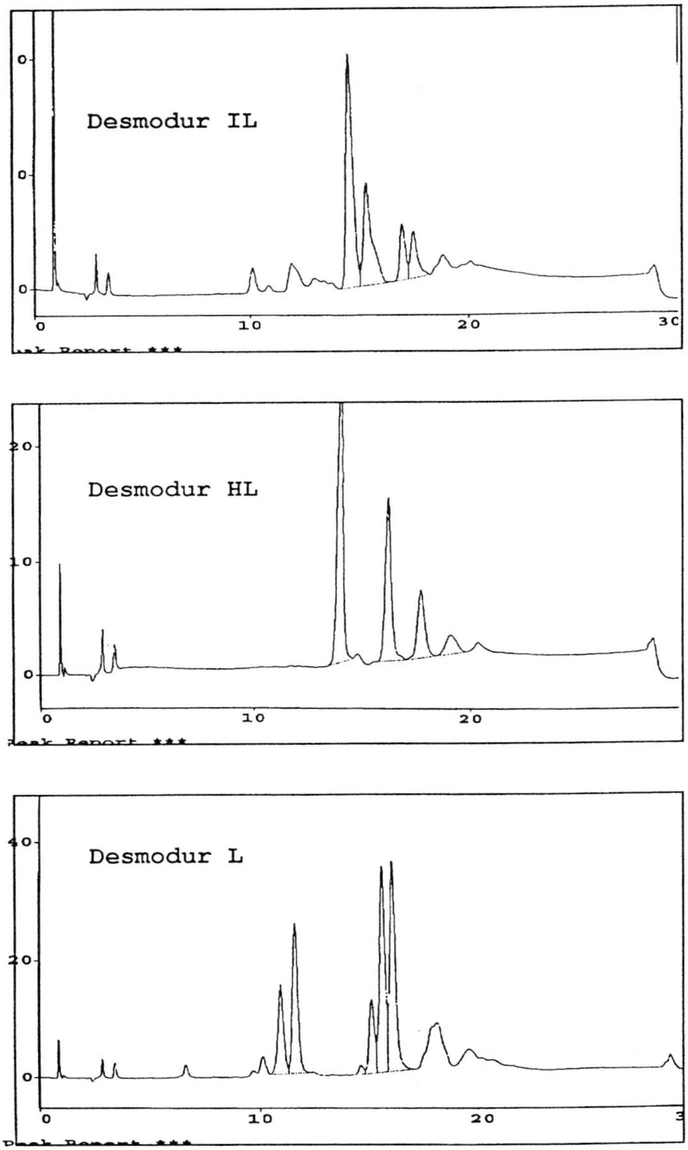

Fig. 1. Example chromatograms for Desmodur L, Desmodur HL and Desmodur IL

Solvent mixtures, Introduction

Method number 1–6

Application Air analysis

Analytical principle Gas chromatography

Completed in June 1997

Introduction

Methods 1–6 describe the determination of solvent mixtures in workplace air by adsorption on solid adsorbents. The methods, which each cover parts of the solvent spectrum, differ in the adsorption agent used and the type of desorption and sample injection technique.
For sampling, measured air volumes from the breathing zone are drawn through adsorption tubes with a sampling pump (active sampling) or passive samplers are used (see Method 5). The adsorbed components are then desorbed in the laboratory with solvents or by thermal desorption (see Method 5). A gas chromatograph equipped with a flame ionisation detector is used for analytical separation and quantification. The quantitative evaluation is carried out with a calibration curve for which the solvent concentrations in the calibration standards are plotted against the peak areas determined with an integrator.

Solvent mixtures

Solvent mixtures are used in various ways; in addition to use as their name implies in industry, they are used mainly for technological reasons as additives in the processing of organic surface coatings (e. g. paints or lacquers), but also for cleaning equipment and textiles.
The evaporation of the solvent is a required effect e. g. in paints, and also influences the quality and finish of the surface. In accordance with the required properties and because they are cheap, solvent mixtures are usually distillation fractions of apolar to polar organic chemicals from chemical processes and therefore contain a wide spectrum of isomers and homologous compounds within their class (primary mixtures).
To achieve certain properties necessary for the application, often empirical mixtures of primary mixtures are used (secondary mixtures) which, for example, can contain rapidly evaporating fractions for initial drying and slowly evaporating fractions so that

the surface dries smooth. As a result of the different evaporation behaviour of the components of such secondary mixtures, the pattern of peaks obtained in the chromatogram is not constant and this makes peak identification difficult.
Classical (primary) solvent mixtures are known under a wide variety of trade names. Depending on the type, they have different properties and contain different proportions of isomeric and homologous compounds. The composition of a secondary mixture is designed to suit the specific application, is often a company secret and therefore is generally not expected to be known. The primary mixtures can be obtained from the chemical trade. Examples of known solvent mixtures are dry cleaning spirit, petroleum ether and Solvesso®. Table 1 shows typical components of some products containing solvents.
The "Solvent mixtures" set of methods, differentiates according to chemical and physical properties (adsorption properties, detectability) and is therefore intended for recognising and analysing primary mixtures.
As solvent mixtures are usually not analysed without some previous knowledge of their composition, and as this previous knowledge is necessary for correct analytical determination, a method can be readily extended to other mixtures with similar properties. In some cases it is advantageous to obtain specimens of the liquid phases used at the sampling site and then in the laboratory to identify the components which they contain (e.g. using GC/MS) before analysing the air samples. In addition, comparison of the patterns of peaks obtained in gas chromatography with those for the substances found in the workplace air can facilitate the recognition of primary mixtures.
Solvents are usually highly to moderately volatile liquids. Vapour released during the production or use of solvents can represent a health hazard at the workplace.
For numerous solvents there are threshold limit values for the workplace air [1, 2]. Generally, however, these apply only for the pure solvent. Simultaneous or consecutive exposure to vapour from different solvents can lead to synergisms.

Table 1. Components of products containing solvents.

Solvent component	Solvent cleanser, degreasing agents	Paints	Glues	Thinners
Alcohols		X	X	X
Petroleum-type hydrocarbons (alkanes), boiling point 180–210 °C	X	X	X	X
Petroleum-type hydrocarbons (aromatics), boiling point 180–210 °C	X	X	X	X
Esters		X	X	X
Ethers			X	
Glycol compounds (glycol ethers, glycol esters)	X	X		X
Halogenated hydrocarbons	X		X	X*
Ketones	X	X	X	X
Terpene hydrocarbons		X		
Others		X		

* (except paint thinners)

To evaluate the exposure situation for a mixture, a substance index is determined for each solvent; this is calculated as a quotient from the concentration measured and the corresponding threshold limit value. These substance indices are added together to give the exposure index. At present substances with MAK values are differentiated from those with TRK values. The threshold limit value is adhered to when the exposure index is smaller than 1 [3].
Many solvent mixtures contain substances for which there is no threshold limit value. If these mixtures contain only hydrocarbons (e.g. aromatics, aliphatics, cycloaliphatics) and are free of additives, the threshold limit value in air is derived from the concentration of aromatics, *n*-hexane, and cyclohexanes/isohexanes in the liquid solvent mixture used. Evaluation of the exposure situation is then carried out in the same way as for solvents with threshold limit values. Threshold limit values are in preparation for fuels and other solvent mixtures which cannot be considered as additive-free hydrocarbon mixtures.

Notes on Methods 1–6

The "Solvent mixture" methods 1–6 cover the following analytical procedures:

1 Adsorption on activated carbon/desorption with carbon disulfide/GC with FID
2 Adsorption on activated carbon/desorption with diethyl ether/GC with FID
3 Adsorption on activated carbon/desorption with a ternary mixture/GC with FID
4 Adsorption on activated carbon/desorption with dimethylformamide, dimethylacetamide, benzyl alcohol or phthalic acid dimethyl ester/head-space GC with FID
5 Adsorption on Tenax TA or XAD-4/thermal desorption/GC with FID
6 Adsorption on silica gel/desorption with water/head-space GC with FID

Methods 1–6 provide for the analysis of different and sometimes overlapping parts of the solvent spectrum.
The six different methods were tested in the ad hoc working group "Solvent Mixtures" for the following substance groups:

Method	Substance groups	Classifications	
1	I	I	Halogenated hydrocarbons, alkanes
2	II, III	II	Glycol compounds
3	II, III	III	Esters/ketones
4	I, III	IV	Alcohols
5	I–III		
6	IV		

Each substance group was tested in collaborative studies with mixtures of different individual substances which are regarded as typical examples of their substance group. For each collaborative study, sampling tubes were loaded by a central laboratory with a

test gas mixture of substances from one substance group. For assessment of the influence of humidity, a test gas with a relative humidity of at least 40 % was used. The test gas concentrations in general are in the lower concentration range, i. e. about 1/10 of the appropriate threshold limit values.
The tests were carried out by at least 2 examiners with 6 tubes in each case, to enable statements to be made on accuracy and precision. The samples were also stored under defined conditions, so that in some cases the methods include information on the shelf-life. In addition the uncertainty associated with the analysis was determined by at least one of the two examiners – as stipulated in DIN EN 482 "Workplace atmosphere – General requirements for the performance of procedures for the measurement of chemical agents" [4] – for each of 6 samples in the low, middle and upper concentration range.

Selection of the analytical method

Table 2 shows the suitability of the six analytical methods for various typical solvent components; for each component only the most suitable method(s) are given or the methods which have been tested with the components.
The components to be determined are listed in the "Substance" column and the suitable analytical procedures for this substance are given in the next columns. It is usually advantageous to use as few different analytical procedures as possible. Otherwise sampling and analysis become very time-consuming. It must also be borne in mind that in addition to the analytes also other unidentified substances can interfere with the loading of the collection phase during sampling.

Table 2. Selection scheme for collection phases.

Substance	1 AC/CS_2	2 AC/ DEE	3 AC/ ternary mixture	4 AC/ HS-GC	5 Tenax or XAD-4	6 SiO_2/ H_2O
Acetone	X	X		X	X	
Benzene	X			X	X	
n-Butyl alcohol	X		X	X	X	X
Isobutyl alcohol	X			X	X	X
2-Butanone	n.s.	n.s.	n.s.	n.s.	X	**X**
Ethylene glycol monobutyl ether			X		X	
Ethylene glycol monobutyl ether acetate	X	X	X	X		
Butylbenzene			X	X	X	
sec-Butylbenzene			X	X	X	
2-Hexanone			X	X		

Table 2. (continued)

Substance	1 AC/CS_2	2 AC/ DEE	3 AC/ ternary mixture	4 AC/ HS-GC	5 Tenax or XAD-4	6 SiO_2/ H_2O
Cyclohexane	X	X		**X**	**X**	
Cyclohexanone		X		X	X	
n-Decane	X	X		X	X	
Diacetone alcohol	X		X			
Dichloromethane	**X**			**X**	**X**	
Diisobutyl ketone		X		X		
Dipropylene glycol monomethyl ether		X			X	
n-Butyl acetate	X	X		**X**	**X**	
Isobutyl acetate	X	X	X	X	X	
Ethyl acetate	X	X	X	X	X	
Methyl acetate	X	X		X	X	
n-propyl acetate	X	X		X	X	
Isopropyl acetate		X		X	X	
Ethanol					X	X
Ethylene glycol monoethyl ether	X		X		X	
Ethylene glycol monoethyl ether acetate	X		X	X	X	
Ethylbenzene	X	X	X	X	X	
2-Ethyl methyl benzene			X	X	X	
3-Ethyl methyl benzene			X	X	X	
4-Ethyl methyl benzene			X	X	X	
n-Heptane	X	X		X	X	
4-Heptanone			X			
n-Hexane	X	X		X	X	
Isopropyl alcohol	X				X	X
Methanol					X	X
Ethylene glycol monomethyl ether			X		X	
Ethylene glycol monomethyl ether acetate	X			X	X	
Propylene glycol 1-methyl ether			X	X	X	
Propylene glycol 1-methyl ether-2-acetate	X	X	X			
Methylcyclohexane		X		X	X	
Methylacrylic acid methyl ester	X				X	
Hexone	X	X	X	X	Xx	
n-Nonane	X	X	X	X	X	
n-Octane	X	X		X	X	
n-Propylbenzene	X	X	X	X	X	
Isopropyl benzene	X	X		X	X	
Styrene	X			X	X	
Tetrachloroethylene	X			X	X	

Table 2. (continued)

Substance	1 AC/CS_2	2 AC/ DEE	3 AC/ ternary mixture	4 AC/ HS-GC	5 Tenax or XAD-4	6 SiO_2/ H_2O
Tetrahydrofuran	X				X	
1,2,3,5-Tetramethyl-benzene			X		X	
Toluene	X	X	X	X	X	
1,1,1-Trichloroethane	X			X	X	
1,1,2-Trichloroethane	X			X	X	
Trichloroethylene	X			X	X	
1,2,3-Trimethyl benzene	X	X	X		X	
1,2,4-Trimethyl benzene	X	X			X	
1,3,5-Trimethyl benzene	X	X	X		X	
o-Xylene	X	X	X	X	X	
m-Xylene	X	X	X	**X**	**X**	
p-Xylene	X	X		X	X	

AC = Activated carbon ; DEE = Diethyl ether;
X = Analytical procedure is suitable for this substance (**bold** indicates methods tested for this substance);
n.s. = not storable (even on storage for a short time, substance losses are to be expected).

References

[1] DFG (1997) MAK- und BAT-Werte-Liste 1997. Senatskommission zur Prüfung gesundheitsschädlicher Arbeitsstoffe, Mitteilung 33. WILEY-VCH-Verlagsgesellschaft, Weinheim
[2] Bundesministerium für Arbeit und Sozialordnung: TRGS 900. Grenzwerte in der Luft am Arbeitsplatz – MAK- und TRK-Werte -. In: Technische Regeln und Richtlinien des BMA zur Verordnung über gefährliche Stoffe. Bundesarbeitsblatt 6/1994
[3] Bundesministerium für Arbeit und Sozialordnung: TRGS 403. Bewertung von Stoffgemischen in der Luft am Arbeitsplatz. In: Technische Regeln und Richtlinien des BMA zur Verordnung über gefährliche Stoffe. Bundesarbeitsblatt 10/1989
[4] Europäisches Komitee für Normung (CEN): DIN EN 482-Arbeitsplatzatmosphäre-Allgemeine Anforderungen an Verfahren für Messung von chemischen Arbeitstoffen. Brüssel 1994. Beuth Verlag, Berlin 1994

Authors: *J. Angerer, E. Flammenkamp, R. Hebisch, W. Krämer, M. Kuck, N. Lichtenstein, U. Risse, U. Schröter, M. Tschickardt*
Examiners: *E. Flammenkamp, K. Goßler, R. Hebisch, M. Hennig, U. Knecht, W. Krämer, M. Kuck, U. Schröter, M. Tschickardt*

Solvent mixtures

Method number	1
Application	Air analysis
Analytical principle	Gas chromatography (charcoal/CS2)
Completed in	June 1997

Summary

The method describes the determination of solvent mixtures in the air at the workplace by adsorption on charcoal and subsequent desorption with carbon disulfide. With a sampling pump, measured air volumes from the breathing zone are drawn through a charcoal tube. In the laboratory the adsorbed components are desorbed with carbon disulfide. A gas chromatograph, equipped with a flame ionisation detector, is used for analytical separation and quantification. Each sample should be analysed with two columns of different polarity. The quantitative evaluation is carried out with a calibration curve for which the solvent concentrations in the calibration standards are plotted against the peak areas determined with an integrator.

Precision of the whole procedure:	Standard deviation (rel.)	s = 1.1–2.2 %
	Mean variation	u = 2.8–5.5 %
	in the concentration range from 0.1 to 2 times the MAK value (for the substances given in Section 8.2) and where n = 6 determinations	
Quantification limit:	0.2–0.8 mg/m^3, referring to a sample volume of 6 L	
Recovery rate:	η > 0.90 (> 90 %) (Section 8.3)	
Sampling recommendations:	Sampling time:	2 hours
	Sample volume:	6 L

Solvent mixtures

See general section on methods 1–6 "Solvent mixtures, Introduction".

Authors: *J. Angerer, E. Flammenkamp, R. Hebisch, W. Krämer, M. Kuck, N. Lichtenstein, U. Risse, U. Schröter, M. Tschickardt*

Examiners: *E. Flammenkamp, K. Goßler, R. Hebisch, M. Hennig, U. Knecht, W. Krämer, M. Kuck, U. Schröter, M. Tschickardt*

Solvent mixtures

Method number 1

Application Air analysis

Analytical principle Gas chromatography (charcoal/CS2)

Completed in June 1997

Contents

1 General principles

The method described here is for the determination of solvent mixtures in the air at the workplace by adsorption on charcoal and subsequent desorption with carbon disulfide. With a sampling pump, measured air volumes from the breathing zone are drawn through a charcoal tube. In the laboratory the adsorbed components are desorbed with carbon disulfide. A gas chromatograph equipped with a flame ionisation detector is used for analytical separation and quantification. Each sample should be analysed with two columns of different polarity. The quantitative evaluation is carried out with a calibration curve for which the solvent concentrations in the calibration standards are plotted against the peak areas determined with an integrator.

2 Equipment, chemicals and solutions

2.1 Equipment

Gas chromatograph equipped with flame ionisation detector, recommended with autosampler
Integrator/computer
Capillary columns with different polarity
Sampling pump for flow rates of about 50 mL/min
Charcoal adsorption tubes (type NIOSH: 100 mg charcoal in the adsorption phase, 50 mg in the control phase)
Tube holders for personal air sampling
Gasmeter or stop clock and soap bubble flowmeter
Thermometer
Hygrometer
Barometer
Various crimp top vials with PTFE-coated septa
Aluminium seal
Seal crimper and decapper
Various volumetric flasks
Various bulb pipettes (e. g. Dispensette)
Various microliter syringes
Analytical balance
Glass cutter
Shaker or ultrasonic bath

2.2 Chemicals

The substances to be determined, analytical grade purity.
Carbon disulfide (benzene-free)

2.3 Calibration standards

Gas chromatographic determination is carried out using a calibration curve.
The procedure for the determination of a solvent mixture made up of benzene, toluene and *m*-xylene is described below.
Stock solution: The stock solution is prepared in a vial with a valve. It is recommended that a sample vial with a capacity of about 15 mL be used. 13 mL carbon disulfide is added to the vial with a Dispensette. 50 µL benzene and 200 µL of each of toluene and *m*-xylene are added with a syringe. It is recommended that the volumes to be dispensed be in the upper third of the dispensing range of the syringes used. All volumetric doses are checked gravimetrically. The values listed in Table 1 were calculated from the respective densities at 20 °C and an assumed purity of 99.7 % for benzene and toluene and of >99.5 % for *m*-xylene.

Table 1. Pipetting scheme for the stock solution.

Substance	Volume dispensed µL	Weight (theoretical) mg	Theoretical concentration µg/µL
Carbon disulfide	13000	16302	
Benzene	50	43.94	3.257
Toluene	200	172.60	12.794
m-Xylene	200	172.80	12.784

The gravimetric weights are used for the calculation of the concentrations.
Calibration standards: The calibration standards are prepared from the stock solution as shown in Table 2.

Table 2. Pipetting scheme for the calibration standards.

Calibration standard	Carbon disulfide		Amount of stock solution to be added	
	Volume µL	Weight (theoretical) mg	Volume µL	Weight (theoretical) mg
1	9995	12618	5	6.25
2	9990	12612	10	12.5
3	9980	12600	20	25.0
4	9960	12574	40	50.0
5	9920	12524	80	100.0
5	9880	12474	120	150.0

Production of the calibration standards is also checked gravimetrically. In the calibration standards the following concentrations are obtained for the substances to be determined:

Table 3. Concentration of the calibration standards.

Calibration standard	Benzene µg/10 mL	Toluene µg/10 mL	*m*-Xylene µg/10 mL
1	16.3	64.0	63.9
2	32.6	127.9	127.8
3	65.1	255.9	255.7
4	130.3	511.8	511.4
5	260.6	1023.5	1022.7
6	390.9	1535.3	1534.1

Each calibration solution is transferred to 2 autosampler vials and analysed chromatographically.
Using the calibration standards described guarantees that after desorption of benzene, toluene and *m*-xylene with 10 mL carbon disulfide from the charcoal, loads of 10–400 µg benzene and 50–1500 µg toluene and *m*-xylene can be determined.

3 Sample collection and preparation

3.1 Sample collection

With a pump equipped with a flow meter or gasmeter, air is drawn with a sampling rate of 50 mL/min through a charcoal tube at a constant flow rate for a maximum sampling time of 2 h. During sampling, the charcoal tube is placed vertically in a suitable tube holder with the inlet at the top and is fastened in the breathing zone of the person or used in a static position at the site of sampling.
The parameters important for determining the concentration, such as sampled air volume, temperature and air pressure, must be determined. The sampling tubes are closed after sampling with the plastic caps provided and labelled clearly, and the sampling data are noted in the sampling protocol. The sampling rate of the pump must also be checked after the end of sampling and recorded. It should not differ by more than 5% from the value originally set [1].
Loaded charcoal tubes should not be stored for more than 14 days and should be kept in the dark. To prevent substance breakthrough, the sampling volume or the sample flow rate should not be greatly exceeded.

3.2 Preparation of the sample

For analytical determination the solvent components adsorbed on the charcoal must be desorbed. This is carried out by adding carbon disulfide. After complete desorption, aliquots of this solution are injected into the gas chromatograph and analysed.

To determine the blank value an unloaded tube must be prepared in the same way as the sample tube.
The charcoal tubes are opened with a glass cutter and the contents are transferred, depending on the tube used, either to a 2 mL vial (150 mg charcoal) or to a 10 mL vial (1 g charcoal). Then either 2 mL or 10 mL carbon disulfide are added and the vials are sealed. Desorption is usually complete after treatment for five minutes in an ultrasonic bath or for half an hour on a shaker or after the samples have been left standing overnight (about 15 hours) (see Table 7). An aliquot of the sample solution is transferred to another sample vial (e.g. an autosampler vial), spiked with an internal standard if necessary and the vial is then closed.

4 Operating conditions for gas chromatography

4.1 Injection

The most common injection technique for investigating highly volatile components is split injection. 1 µL of the desorption solution is injected into the gas chromatograph. The injector is usually kept at a temperature between 200 °C and 250 °C for the desorption solvents used and the analytes determined in the method described here. The given detection limits are usually achieved with a split ratio of 1:20. To improve the detection limits, splitless injection into the column can be used if the then wider desorption agent peak can still be separated from the substance peaks.

4.2 Chromatography

If there are only a few solvent components in the air at the workplace the relatively low separation potential of packed columns may be sufficient for gas chromatographic separation of the individual substances. Capillary columns are usually used to solve complex separation problems. As a result of their better resolving power and less peak tailing, capillary columns produce better separations. Depending on the classes of compound to be separated, capillary columns with non-polar separation phases, such as DB-1, SPB-1, OV-1, SE-30 (dimethylpolysiloxane), SE-52 (95% dimethylpolysiloxane and 5% phenylpolysiloxane), moderately polar phases, such as OV 17 (50% dimethylpolysiloxane and 50% phenylpolysiloxane), OV-1701 (86% dimethylpolysiloxane and 14% cyanopropylphenylpolysiloxane) or polar phases, such as Carbowax, DB-WAX (polyethylene glycol) are most suitable for separation.
With a non-polar column 2-methylheptane, for example, can serve as internal standard, with a polar column *n*-undecane. For analyses of airborne samples, capillary columns with lengths of 30 m, internal diameters of 0.25 to 0.5 mm and film thickness of 0.5 µm and 1 µm have proved reliable in practice.
To determine compounds with greatly different boiling points and to minimise the time needed for separation, separation is optimised by using temperature programmes.

As a non-specific detector such as the flame ionisation detector (FID) is usually used, the result should be checked by separation on two capillary columns of different polarity. This can be carried out by successive injection into two capillary columns of different polarity or by simultaneous injection into two capillaries (dual capillary GC) connected to the one injector. A substance is regarded as having been identified when the concentration determined after gas chromatographic separation has the same value on both capillaries. If the analytical results for an identified substance differ significantly on the two columns, there is probably overlapping of peaks on one of the two columns. If this overlapping cannot be remedied by changing the temperature programme, the smaller value should generally be taken when calculating the analytical result.
The GC conditions can also be adapted to fit individual problems. For example, a second column is not necessary if all components are known and no overlapping occurs.
Below are examples of two sets of separation conditions for analysing various solvent mixtures.

4.2.1 Method A

Columns:	DB-5 or DB-1701 (length 30 m, internal diameter 0.25 mm, film thickness 1 µm)
Temperature programme:	40 °C; 9 min; 5 °C/min up to 250 °C; 250 °C constant
Sample injection:	on-column
Carrier gas:	Helium (140 kPa column pressure)
Detector gas (FID):	Hydrogen (60 kPa, 20 mL/min)
	Synthetic air (110 kPa, 32 mL/min)
	Helium (make-up gas, 150 kPa, 40 mL/min)

4.2.2 Method B

Column:	SPB-1 (length 30 m, internal diameter 0.25 mm, film thickness 1 µm)
Temperature programme:	40 °C; 2 min; 20 °C/min up to 180 °C; 180 °C constant
Sample injection:	Split ratio 1: 20
Carrier gas:	Helium (70 kPa column pressure)
Detector gas (FID):	Hydrogen (60 kPa, 20 mL/min)
	Synthetic air (110 kPa, 32 mL/min)
	Helium (make-up gas, 150 kPa, 40 mL/min)

Figure 1 shows an example of a chromatogram for a solvent mixture after desorption with carbon disulfide.

5 Analytical determination

The gas chromatograph is set up as given in the operating conditions in Section 4. 1 µL from each sample is injected into the gas chromatograph. Control solutions of known content are analysed at regular intervals between the samples to check the calibration. Each sample should be injected into the chromatograph at least twice.
Evaluation is carried out according to the equipment available using a suitable peak area integration system.

6 Calibration

When preparing the calibration curve, each calibration standard is injected into the gas chromatograph twice. Two autosampler vials are filled with each calibration standard (see Section 2.3) and the solutions are analysed as described. The calibration curves are drawn up from the amounts of the substance analysed and the peak area values obtained. In routine analysis the calibration curve must be checked regularly.

7 Calculation of the analytical result

On the basis of the peak areas obtained, the weights of the individual components are read from the appropriate calibration curves. The concentrations by weight ρ are calculated according to the following equation:

$$\rho = \frac{X}{V \cdot \eta} \cdot \frac{273 + t_g}{273 + t_a} \tag{1}$$

At 20 °C and 1013 hPa:

$$\rho_0 = \rho \cdot \frac{273 + t_a}{293} \cdot \frac{1013 \text{ hPa}}{p_a} \tag{2}$$

where:

ρ is the concentration by weight of a component in mg/m^3

ρ_0 is the concentration by weight in mg/m^3 at 20 °C and 1013 hPa

X is the weight of the component in the solution in µg

V is the sampled air volume in L (determined with a gasmeter or calculated from the duration of sampling and flow rate)

η is the recovery rate (for calibrations where the calibration standards are spiked with adsorption agent before analysis, η does not need to be determined. In these cases 1 is substituted for ηin the above equation.)

t_g is the temperature in the gasmeter in °C
t_a is the temperature of the ambient air in °C
p_a is the atmospheric pressure of the ambient air in hPa

8 Reliability of the method

Validation of the method was carried out with air concentrations equivalent to the MAK values valid in 1995. At present (2000) the following MAK values apply: cyclohexane: 700 mg/m^3; dichloromethane: carcinogen category 3A; 1,1,1–trichloroethane: 1100 mg/m^3 [2].

8.1 Accuracy

Comparative experiments were carried out using a solvent mixture to test the accuracy of the method. In a laboratory a dynamic test gas was produced for this purpose using continuous injection [3].
The sampling tubes were each loaded for 2 hours with about 6 L of the test gas (40% relative humidity at 20 °C). After loading in parallel, the tubes were sent to a number of laboratories and analysed there according to the method described here. The results for various substances are listed in Table 4.

Table 4. Accuracy tests.

Substance	Load	Found			
		Examiner I		Examiner II	
	mg/m^3	mg/m^3	%	mg/m^3	%
n-Hexane	18.5	17.7	95.7	18.3	98.9
Cyclohexane	105	99.7	95.1	102	97.1
Toluene	31.7	29.9	94.4	30.9	97.5
Dichloromethane	34.8	33.3	95.7	36.3	104
1,1,1-Trichloroethane	107	102	94.9	104	96.7

8.2 Precision

In the comparative experiments described in Section 8.1 also the relative standard deviations (including sampling) were determined for 6 parallel samples.

Table 5. Standard deviation (relative) s, $n = 6$ determinations.

Substance	Load	Standard deviation (rel.) s	
		Examiner I	Examiner II
	mg/m^3	%	%
n-Hexane	18.5	0.43	1.6
Cyclohexane	105	0.44	1.5
Toluene	31.7	0.68	1.5
Dichloromethane	34.8	1.1	1.3
1,1,1-Trichloroethane	107	0.45	1.3

To determine the uncertainty associated with the analysis in the range in question, the relative standard deviations s of the method in the low (approx. 0.1 × MAK), middle (approx. 1 × MAK) and upper (approx. 2 × MAK) concentration range were determined by an examiner according to EN 482 [4]. With a dynamic test gas apparatus three test gases of different concentrations were produced and each of 10 charcoal tubes were loaded for 120 minutes with a flow rate of 50 mL/min.

Table 6. Standard deviation (rel.) s for different concentrations, $n = 10$ determinations.

Substance	MAK value mg/m^3	Conc. mg/m^3	s %	Conc. mg/m^3	s %	Conc. mg/m^3	s %
n-Hexane	180	19.1	1.41	162.4	1.71	323.3	1.30
Cyclohexane	1050	114.5	1.16	–	–	–	–
Toluene	190	36.1	1.39	369.0	1.22	736.1	1.92
Dichloromethane	360	36.7	1.47	340.6	1.15	679.0	2.17
1,1,1-Trichloroethane	1080	115.6	1.33	979.0	1.05	1952.1	1.33

As substance breakthrough can occur with the higher cyclohexane concentrations given when other substances are present and sampling tubes with 150 mg charcoal are used as adsorption agent, the standard deviation at these concentrations was not determined.

8.3 Recovery rate

The recovery rate for alkanes and chlorinated hydrocarbons is usually in the range of 0.95 to 1.

During the validation of the method the percentage desorption of the substances investigated was determined by examiner 1 according to the phase equilibrium procedure (at about 0.1 times the MAK value) and by examiner II using test gas in the low, middle and upper concentration range.

Table 7. Desorptions [%] with different concentrations.

Substance	Desorption [%]			
	Examiner I		Examiner II	
	0.1 × MAK	0.1 × MAK	1 × MAK	2 × MAK
n-Hexane	101	94.1	95.4	92.9
Cyclohexane	102	92.2	–	–
Toluene	97.7	89.3	91.0	88.6
Dichloromethane	98.7	94.4	94.2	90.0
1,1,1-Trichloroethane	102	93.0	95.2	92.7

8.4 Quantification limit

Quantification limits between 0.2 and 0.8 mg/m^3 were determined for the substances listed.

8.5 Specificity

As a result of the low specificity of flame ionisation detectors, interference from components with the same retention time is possible with some gas chromatographic procedures. To check the analytical results, as described above, a second column of different polarity can be used as well, or possibly also a mass spectrometric detector.

9 Discussion of the method

The practicability of this procedure has been demonstrated in numerous analyses to determine BTX aromatics and halogenated hydrocarbons. Participation in intercomparison programmes (including sampling) has verified the accuracy of the analytical results.
Breakthrough of solvents can occur if the sampling time is long or the concentrations are very high. To recognise breakthrough, the two zones of the charcoal tubes can be analysed separately or instead sampling tubes with a greater amount of charcoal (e. g. 1 g charcoal/desorption with 10 mL CS_2) can be used immediately.
If relevant unknown components are detected in a sample, these must first be identified using GC/MS analysis.

10 References

[1] Europäisches Komitee für Normung (CEN): DIN EN 1232-Arbeitsplatzatmosphäre – Pumpen für die personenbezogene Probenahme von chemischen Stoffen – Anforderungen und Prüfverfahren. Brüssel 1993. Beuth Verlag, Berlin 1993.
[2] DFG (2000) MAK- und BAT-Werte-Liste 2000. Senatskommission zur Prüfung gesundheitsschädlicher Arbeitsstoffe, Mitteilung 36. Wiley-VCH Verlag, Weinheim.
[3] Verein Deutscher Ingenieure (VDI): VDI-Richtlinie 3490-Prüfgase, Blatt 1–16, Beuth Verlag, Berlin.
[4] Europäisches Komitee für Normung (CEN): DIN EN 482-Arbeitsplatzatmosphäre-Allgemeine Anforderungen an Verfahren für Messung von chemischen Arbeitstoffen. Brüssel 1994. Beuth Verlag, Berlin 1994.

Authors: *J. Angerer, E. Flammenkamp, R. Hebisch, W. Krämer, M. Kuck, N. Lichtenstein, U. Risse, U. Schröter, M. Tschickardt*
Examiners: *E. Flammenkamp, K. Goßler, R. Hebisch, M. Hennig, U. Knecht, W. Krämer, M. Kuck, U. Schröter, M. Tschickardt*

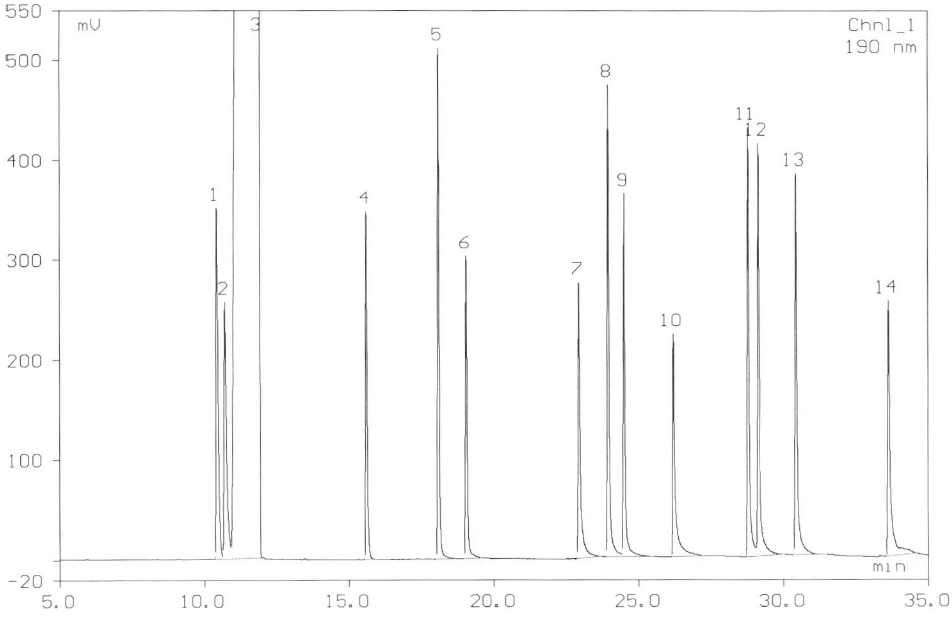

Fig. 1. Example of a chromatogram for a solvent mixture after desorption with carbon disulfide (chromatographic conditions see Section 4.2.1 method A; **1**: acetone, **2**: *iso*-propanol, **3**: carbon disulfide, **4**: ethyl acetate, **5**: benzene, **6**: *n*-heptane, **7**: 4-methylpentan-2-one, **8**: toluene, **9**: *n*-octane, **10**: *n*-butyl acetate, **11**: ethyl benzene, **12**: *m*-xylene, **13**: *p*-xylene, **14**: 1,3,5-trimethyl benzene).

Solvent mixtures

Method number 2

Application Air analysis

Analytical principle Gas chromatography (activated carbon/diethyl ether)

Completed in June 1997

Summary

The method described here is for the analysis of solvent mixtures in workplace air by enrichment on activated carbon and subsequent desorption with diethyl ether. With a sampling pump, measured air volumes from the breathing zone are drawn through sampling tubes filled with activated carbon (type B, Dräger). The adsorbed components are eluted with diethyl ether. A gas chromatograph equipped with a flame ionisation detector is used for analysis. Each sample is analysed with two columns of different polarity. Quantitative evaluation is carried out using multiple-point calibration with an internal standard.

Precision of the whole procedure:	Standard deviation (rel.)	s = 2.0–8.7%
	Mean variation	u = 5–22%
	in the concentration range from 0.1 to 2 times the MAK value (for the substances given in Section 8.2), n = 6 determinations	
Limit of quantification:	0.3–1.5 mg/m^3, for a sampled air volume of 40 L (for the substances given in Table 7)	
Recovery:	η = 0.65–0.98 (65–98%) (see Tab. 7)	
Sampling recommendation:	Sampling time:	2 h (8 h)
	Sample volume:	40 L

Solvent mixtures

See general section on methods 1–6 "Solvent mixtures, Introduction"

Authors: *J. Angerer, E. Flammenkamp, R. Hebisch, W. Krämer, M. Kuck, N. Lichtenstein, U. Risse, U. Schröter, M. Tschickardt*
Examiners: *E. Flammenkamp, K. Goßler, R. Hebisch, M. Hennig, U. Knecht, W. Krämer, M. Kuck, U. Schröter, M. Tschickardt*

Solvent mixtures

Method number	2
Application	Air analysis
Analytical principle	Gas chromatography (activated carbon/diethyl ether)
Completed in	June 1997

Contents

1 General principles

With a sampling pump, measured air volumes from the breathing zone are drawn through adsorption tubes filled with activated carbon (type B, Dräger). The adsorbed components are eluted with diethyl ether. A gas chromatograph equipped with a flame ionisation detector is used for analysis. Each sample should be analysed with two columns of different polarity. Quantitative evaluation is carried out using multiple-point calibration with an internal standard.

2 Equipment, chemicals and solutions

2.1 Equipment

Gas chromatograph equipped with flame ionisation detector, possibly with autosampler
Integrator/computers
Capillary columns with different polarity
Pump with flow rate adjustable to 4–20 L/h
Sampling tubes (e. g. Dräger activated carbon tubes type B: 300/700 mg activated carbon)
Gasmeter or stop clock and soap bubble flowmeter
Thermometer
Hygrometer
Barometer
Vials with PTFE-coated septa
Aluminium crimp caps
Crimping tongs for sealing and opening the vials
10 mL Volumetric flasks
10 mL Bulb pipettes
10, 250, 500 and 1000 μL Syringes

2.2 Chemicals

The substances to be determined as shown in Table 7, analytical grade purity
Diethyl ether, analytical grade
2-Methylheptane for gas chromatography (as internal standard for non-polar columns)
n-Undecane for gas chromatography (as internal standard for polar columns)

2.3 Calibration standards

Calibration standards are required for drawing up the necessary calibration curves.
The calibration standards should cover at least the range from 0.1 to 2 times the threshold limit value (TLV). If this is not possible because the curve is not linear, calibration

is carried out within the linear range and the samples are diluted accordingly. Below the procedure for the determination of ethyl acetate, isobutyl acetate and hexone is described.

Stock solution: The stock solution is prepared in a 10 mL vial with a valve. 10 mL diethyl ether is added to the vial with a bulb pipette. 500 µL of each of the substances to be calibrated, ethyl acetate, isobutyl acetate and hexone, is added with a syringe. It is recommended that the volumes to be dispensed be in the upper third of the dispensing range of the syringes used. All volumetric doses are checked gravimetrically. The values listed in Table 7 were calculated from the respective densities at 20 °C and an assumed purity of 99.5% for ethyl acetate, of >98% for isobutyl acetate and of 99% for hexone.

Table 1. Pipetting scheme for the stock solution.

Substance	Volume dispensed µL	Sample weight (theoretical) mg	Theoretical concentration µg/µL
Diethyl ether	10000		
Ethyl acetate	500	451.39	39.15
Isobutyl acetate	500	438.58	38.03
Hexone	500	399.66	34.83

The gravimetric control should yield deviations from the theoretical value of less than 1%. If the deviation is more than 2%, the stock solution must be prepared again.

Calibration standards: The calibration standards are prepared from the stock solution as shown in Table 2. The volumes of the stock solution listed in the table are added to 10 mL volumetric flasks containing diethyl ether and the flasks are then filled to the mark with diethyl ether. 5 µL of a mixture of 2-methyl heptane and n-undecane (equal volume parts of each component) is added to each calibration standard as internal standard. In the calibration solutions the following concentrations are obtained for the substances to be determined.

Table 2. Pipetting scheme for the calibration standards.

Calibration solution	Volume stock solution µL	Ethyl acetate mg/10 mL	Isobutyl acetate mg/10 mL	Hexone mg/10 mL
1	4	0.16	0.15	0.14
2	120	4.70	4.56	4.18
3	240	9.40	9.13	8.36
4	360	14.09	13.69	12.54
5	480	18.79	18.25	16.72
6	600	23.49	22.82	20.90

Each calibration standard is transferred to autosampler vials and analysed chromatographically.
Use of the calibration standards described guarantees the determination of the three components in the range from about 4 mg/m^3 to 550 mg/m^3 or 500 mg/m^3 for hexone.

3 Sample collection and preparation

3.1 Sample collection

With a pump with adjustable flow rate or equipped with a gasmeter, air is drawn through an activated carbon tube (type B, Dräger) at a flow rate of 20 L/h (sampling time max. 2 h). If sample collection is to be carried out over a longer period of time (up to 8 hours), the flow rate should be reduced to 4 L/h. Observation of these parameters guarantees that breakthrough does not occur for the substances named in the method in concentrations up to 4 times their threshold limit value (relative humidity 80%).
During sampling, the activated carbon tube is placed vertically in a suitable tube holder with the inlet at the top in the breathing zone of the person or in a static position. The parameters important for determining the concentration, such as sampled air volume, temperature and air pressure, must be determined. The sampling tubes are closed after sampling with the plastic caps provided and labelled clearly, and the sampling data and data on the work area/enterprise are noted in the sampling protocol.
The output of the pump must also be checked and recorded after the end of sampling. It should not differ by more than 5% from the value originally set. Loaded sampling phases can be stored in the dark for at least 14 days without losses.

3.2 Preparation of the sample

The activated carbon tubes are eluted against the direction of flow with 10 mL diethyl ether. The end of the activated carbon tube which was connected to the pump is attached to a suitable funnel through which 10 mL diethyl ether is added. The eluate is collected in a 10 mL volumetric flask. After elution, the flask is filled to the 10 mL mark with diethyl ether and 5 µL internal standard (2-methylheptane: n-undecane = 1 : 1) is added. If necessary the samples are transferred to autosampler vials.
To ensure that the diethyl ether and activated carbon used do not contain interfering impurities, an unused activated carbon tube from the batch used for sampling is analysed as blank.

4 Operating conditions for gas chromatography

Each sample should be analysed with two columns of different polarity. For the non-polar column 2-methylheptane serves as internal standard, for the polar column *n*-undecane. Overlapping with the components listed in Table 7 does not occur with these standards with the suggested temperature programme.

Columns:	60 m DB-5 and 60 m DB-WAX	
Internal diameter:	0.25 mm	
Film thickness:	0.5 µm	
Temperatures:	Injector:	180 °C
	Detector:	250 °C
	Column thermostat:	8 minutes at 50 °C, 5 °C/min up to 80 °C 5 minutes at 80 °C, 5 °C/min up to 180 °C 6 minutes at 180 °C

The total run time of the temperature programme is 45 minutes. Under these gas chromatographic operating conditions, many common solvent mixtures with different boiling ranges can be analysed. The GC conditions can also be adapted to fit specific problems. For example, a second column (DB-WAX) may not be necessary if all components are known and it is ensured that no overlapping occurs.
Figures 1 and 2 show example chromatograms of a solvent mixture.

5 Analytical determination

1 µL of each sample is injected into the gas chromatograph. Aliquots of the calibration standards are analysed at regular intervals between the samples to check the calibration. Each sample should be injected into the chromatograph at least twice. If only one GC column is used (e.g. DB-5), the number of injections must be increased accordingly.

6 Calibration

Calibration is carried out using an internal standard. Each calibration standard (see Section 2.3) is injected into the gas chromatograph three times and analysed like the sample solution. The injection volume is 1 µL.
To draw the calibration curves, the peak areas are determined using an integration system and plotted against the corresponding substance concentrations. The calibration

curves are linear in the given concentration ranges for the substances tested. Calibration curves should be checked regularly, e. g. during routine analysis by checking at least one point. Blank values must be taken into consideration.

7 Calculation of the analytical result

On the basis of the peak areas obtained, the weights of the individual components are read from the appropriate calibration curves. The concentrations by weight ρ are calculated according to the following equation:

$$\rho = \frac{X}{V \cdot \eta} \cdot \frac{273 + t_g}{273 + t_a} \quad (1)$$

At 20 °C and 1013 hPa:

$$\rho_0 = \rho \cdot \frac{273 + t_a}{293} \cdot \frac{1013 \text{ hPa}}{p_a} \quad (2)$$

where:

ρ is the concentration by weight in mg/m^3
ρ_0 is the concentration by weight in mg/m^3 at 20 °C and 1013 hPa
X is the weight of the component in the solution in µg
V is the sampled air volume (determined with a gasmeter or calculated from the duration of sampling and flow rate) in L
η is the recovery
t_g is the temperature in the gasmeter in °C
t_a is the temperature of the ambient air in °C
p_a is the atmospheric pressure of the ambient air in hPa

8 Reliability of the method

Validation of the method was carried out with air concentrations equivalent to the MAK values valid in 1995. At present (2001) the following MAK values apply: ethyl acetate: 1500 mg/m^3; isobutyl acetate: 480 mg/m^3; 2-butanone: 600 mg/m^3; hexone: 83 mg/m^3 [1].

8.1 Accuracy

Comparative experiments were carried out using a solvent mixture to test the accuracy of the method. For this purpose in a laboratory a dynamic test gas was produced by continuous injection. Sampling tubes were each loaded for 2 hours with up to 40 litres of the test gas at 40% relative humidity ($n = 6$) and in dry air ($n = 2$). After loading in parallel, the tubes were sent to a number of laboratories and analysed there according to this method. The results are shown in Table 3.

Table 3. Accuracy tests.

Substance	Load	Found			
		Examiner I		Examiner II	
	mg/m^3	mg/m^3	%	mg/m^3	%
Ethyl acetate	136	128	94.2	121	89.3
Isobutyl acetate	80.5	72.9	90.6	72.2	89.7
m-Xylene	40.3	35.0	86.8	43.0	107
2-Butanone	19.5			4.6	23.7
Hexone	45.7	44.6	97.5	44.8	98.0

The method was not found to be wholly suitable for 2-butanone as significant substance losses occur when the sample air is moist and the samples are stored for more than one day.

8.2 Precision

In the comparative experiments described in Section 8.1 also the relative standard deviations and mean variations were determined (from n = 6 parallel samples, including sampling).
The values listed in Table 4 were obtained.

Table 4. Standard deviation (relative) s, $n = 6$ determinations.

Substance	Load	Standard deviation (rel) s	
		Examiner I	Examiner II
	mg/m^3	%	%
Ethyl acetate	136	6.6	2.4
Isobutyl acetate	80.5	5.0	2.6
m-Xylene	40.3	12.7	2.7
2-Butanone	19.5		7.2
Hexone	45.7	9.4	2.6

To determine the uncertainty associated with the analysis in the range in question, the relative standard deviation *s* of the method (Table 5) in the low, middle and upper concentration range was determined by an examiner according to EN 482. Using test gas, each of 6 activated carbon tubes were loaded with a solvent mixture.

Table 5. Standard deviation (rel.) *s* for different concentrations, *n* = 6 determinations.

Substance	MAK value mg/m^3	Conc. mg/m^3	*s* %	Conc. mg/m^3	*s* %	Conc. mg/m^3	*s* %
Ethyl acetate	1400	135	3.3	1350	2.0	2814	5.7
Isobutyl acetate*	950	98	3.2	980	2.1	1968	5.3
m-Xylene*	440	43	8.7	430	3.8	866	5.5
2-Butanone	590	60	4.3	600	2.4	1208	6.1
Hexone	400	40	4.4	400	2.3	801	4.9

* (MAK value applies for all isomers)

To determine the precision of the procedure, in addition each of six activated carbon tubes were loaded using test gases with a concentration of various individual components and prepared and analysed as described. The characteristic data determined for the method, such as standard deviations, are given with the corresponding concentrations in Table 7.

8.3 Recovery rate

During comparative investigations, the percentage desorption (Table 6) of the individual substances was determined by examiner I using test gas in the low, middle and upper concentration range and by examiner II according to the phase equilibrium procedure at 0.1 times the MAK value.

Table 6. Desorption [%] with different concentrations.

Substance	Desorption [%]			
	Examiner I			Examiner II
	0.1 MAK	1 MAK	2 MAK	0.1 MAK
Ethyl acetate	81.6	87.0	92.7	96
Isobutyl acetate	87.2	85.7	93.7	98
m-Xylene	73.4	72.6	81.9	67
2-Butanone	79.1	81.2	87.8	93
Hexone	89.3	84.3	91.8	98

In addition, during the validation of the method, the components, in a solution of diethyl ether or undiluted, were added individually to a gas collection vessel with septum and two adjacent connection ports to determine the recovery. The gas collection vessel was connected directly to the activated carbon tube via one port. The air containing solvent was drawn out of the vessel and through the sampling tubes with a pump (20 L/h, 2 hours sampling time). After 1 hour the gas collection vessel was heated, to release any components adsorbed on the glass walls. The recovery was calculated at various concentrations from each of six determinations (analytically determined weight/loaded weight). Table 7 gives examples of values for some substances determined during validation of the method.

Table 7. Characteristic data for the method of desorption with diethyl ether.

Substance	Standard deviation (relative) s^* %			Recovery	Quantification limit mg/m^3
Acetone	8.8 (10)	11.6 (527)	1.4 (2000)	0.65–0.90	0.5
Ethylene glycol monobutyl ether acetate	5.0 (1)	1.2 (64)	3.0 (300)	0.90	0.5
Cyclohexane	2.3 (9)	2.4 (100)	1.7 (4300)	0.93	0.35
Cyclohexanone	1.4 (11.8)	3.8 (1041)		0.76	1.0
n-Decane	7.6 (8)	1.1 (3970)		0.98	0.5
Diisobutyl ketone	2.2 (2)	1.2 (10)	3.5 (886)	0.94	1.0
Dipropylene glycol monomethyl ether	2.3 (20)	2.4 (600)		0.90	1.5
n-Butyl acetate	3.0 (4)	4.2 (483)	0.5 (2000)	0.97	1.3
Isobutyl acetate	2.2 (4)	4.8 (478)	0.8 (2000)	0.96	1.5
Ethyl acetate	3.5 (4)	5.9 (492)	2.3 (2800)	0.95	1.2
Methyl acetate	5.6 (4)	6.7 (510)	1.7 (1200)	0.90	1.0
n-Propyl acetate	2.0 (4)	4.8 (485)	0.5 (2000)	0.95	1.0
Isopropyl acetate	3.0 (4)	6.4 (477)	0.6 (2000)	0.95	1.0
Ethyl benzene	3.4 (5)	5.3 (200)	1.1 (1050)	0.92	0.5
n-Heptane	2.0 (7.8)	2.1 (80)	4.6 (3674)	0.95	0.5
n-Hexane	2.8 (7.5)	2.3 (78)	1.6 (3500)	0.91	0.5
Propylene glycol 1-methyl ether	4.9 (7.6)	4.7 (11.5)	1.4 (235)	0.98	5
[Propylene glycol 1-methyl ether-2]-acetate	4.3 (1)	2.0 (10)	0.7 (450)	0.89	5
Methylcyclohexane	2.1 (8.7)	2.1 (90)	1.9 (4171)	0.95	1
Hexone	2.5 (10)	4.6 (55)	4.4 (2027)	0.89	1.1
n-Nonane	5.9 (8)	1.4 (84)	3.7 (3900)	0.98	0.5
n-Octane	3.6 (8)	1.6 (82)	2.1 (3800)	0.97	0.5
n-Propyl benzene	3.8 (5)	5.3 (200)	4.0 (1000)	0.91	0.5
Isopropyl benzene	3.2 (5)	5.2 (200)	1.9 (1000)	0.94	0.5
Toluene	2.4 (5)	5.7 (200)	1.8 (1000)	0.84	0.3
1,2,3-Trimethyl benzene	5.7 (5)	6.1 (200)	0.8 (1000)	0.70	0.5
1,2,4-Trimethyl benzene	5.2 (5)	5.7 (200)	0.8 (1000)	0.79	0.5

Table 7. (continued)

Substance	Standard deviation (relative) s* %			Recovery	Quantification limit mg/m³
1,3,5-Trimethyl benzene	4.2 (5)	5.5 (200)	0.9 (1000)	0.88	0.5
o-Xylene	5.0 (5)	5.4 (200)	1.7 (1100)	0.77	0.5
m-Xylene	5.4 (5)	5.6 (200)	2.9 (1100)	0.85	0.5
p-Xylene	4.0 (5)	5.3 (200)	2.1 (1100)	0.87	0.5

* The values in brackets are the concentrations in mg/m³, for which the standard deviations were determined.

8.4 Quantification limit

The quantification limits, for a sampled air volume of 40 L, are in the range from 0.3 to 1.5 mg/m³.

8.5 Specificity

As a result of the low specificity of flame ionisation detectors, interferences from components with the same retention time are possible with some gas chromatographic procedures. To check the analytical results, as described above, a second column of different polarity can be used as well, or possibly also a mass spectrometric detector.

9 Discussion of the method

The analytical procedure described is particularly suitable for determining polar components such as esters, alcohols or ketones. With some aromatic compounds lower recovery and higher standard deviations must be expected.

10 References

[1] Deutsche Forschungsgemeinschaft. List of MAK and BAT Values 2001. Commission for the Investigation of Health Hazards of Chemical Compounds, Report No. 37. WILEY-VCH-Verlag, Weinheim, Germany

Authors: *J. Angerer, E. Flammenkamp, R. Hebisch, W. Krämer, M. Kuck, N. Lichtenstein, U. Risse, U. Schröter, M. Tschickardt*

Examiners: *E. Flammenkamp, K. Goßler, R. Hebisch, M. Hennig, U. Knecht, W. Krämer, M. Kuck, U. Schröter, M. Tschickardt*

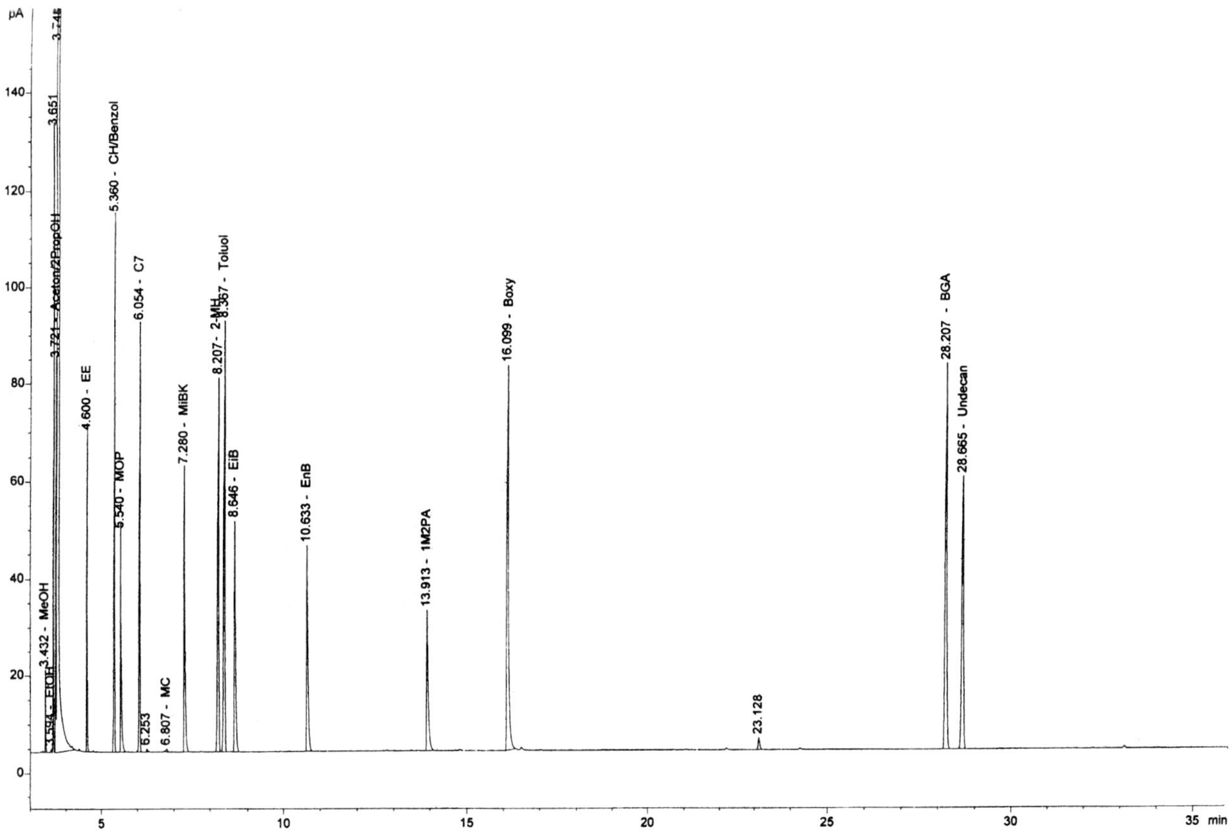

Fig. 1. Example of a chromatogram for a solvent mixture, column: DB-5 (chromatographic conditions see Section 4; EE: ethyl acetate, CH: cyclohexane, MOP: propylene glycol 1-methyl ether, C 7: heptane, MiBK: hexone, 2-MH: 2-methylheptene, EiB: isobutyl acetate, EnB: *n*-butyl acetate, 1M2PA: propylene glycol 1-methyl ether-2-acetate, Boxy: 2-butoxy ethanol, BGA: 2-butoxyethyl acetate).

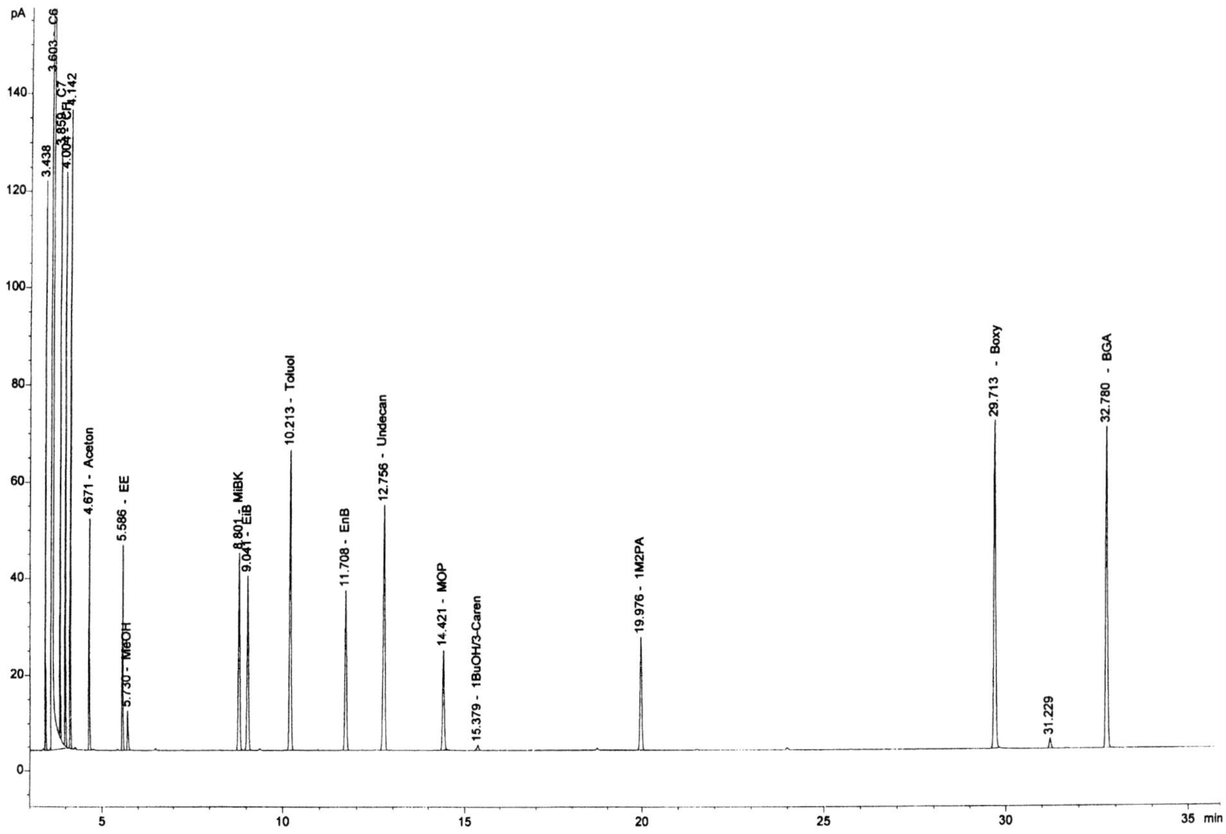

Fig. 2. Example of a chromatogram for a solvent mixture, column: DB-WAX MOP: propylene glycol 1-methyl ether, C 7: heptane, MiBK: hexone, 2-MH: 2-methylheptene, EiB: isobutyl acetate, EnB: *n*-butyl acetate, 1M2PA: propylene glycol 1-methyl ether-2-acetate, Boxy: 2-butoxy ethanol, BGA: 2-butoxyethyl acetate).

Solvent mixtures

Method number 3

Application Air analysis

Analytical principle Gas chromatography (charcoal/ternary mixture)

Completed in June 1997

Summary

The method described here is for the determination of glycol ethers in the air at the workplace by adsorption on charcoal and subsequent desorption with a suitable solvent mixture. In addition, other components of technical grade solvents can also be detected.
The solvent vapours in the air, in particular glycol ethers, are adsorbed on the charcoal of a personal air sampling system. The charcoal is eluted with a suitable desorption agent mixture and the eluate is analysed with a capillary gas chromatograph equipped with a flame ionisation detector. The quantitative evaluation is carried out with calibration curves for which the solvent concentrations in the calibration standards are plotted against the peak areas determined with an integrator.

Precision of the whole procedure:	Standard deviation (rel.)	s = 1.0–9%
	Mean variation	u = 2.6–23%
	for the concentrations given in Section 8.2 and where n = 6 determinations	
Quantification limit:	0.4–1.1 mg/m^3, for a sampled air volume of 28 L (for the compounds given in Table 7)	
Recovery rate:	η = 0.95–1.11 (95–111%) (see Section 8.3)	
Sampling recommendations:	Sampling time:	8 hours
	Sample volume:	<30 L

Solvent mixtures

See general section on methods 1–6 "Solvent mixtures, Introduction".

Authors: *J. Angerer, E. Flammenkamp, R. Hebisch, W. Krämer, M. Kuck, N. Lichtenstein, U. Risse, U. Schröter, M. Tschickardt*
Examiners: *E. Flammenkamp, K. Goßler, R. Hebisch, M. Hennig, U. Knecht, W. Krämer, M. Kuck, U. Schröter, M. Tschickardt*

Solvent mixtures

Method number 3

Application Air analysis

Analytical principle Gas chromatography (charcoal/ternary mixture)

Completed in June 1997

Contents

1 General principles

The solvent vapours in the air, in particular glycol ether vapours, are adsorbed on the charcoal of a sampling system using personal air sampling. The charcoal is eluted using a suitable desorption mixture and the eluate is analysed using a capillary gas chromatograph equipped with a flame ionisation detector. The quantitative evaluation is carried out with calibration curves for which the solvent concentrations in the calibration standards are plotted against the peak areas determined with an integrator.

2 Equipment, chemicals and solutions

2.1 Equipment

Gas chromatograph equipped with flame ionisation detector, possibly with autosampler
Integrator/computer
Sampling tubes (e. g. charcoal tubes, type NIOSH)
Gasmeter or stop clock and soap bubble flowmeter
Thermometer
Hygrometer
Barometer
100, 500, 1000 µL and 5 mL syringes, e. g. from Hamilton
10, 20 and 100 mL volumetric flasks
10 and 100 mL graduated cylinders
1 and 5 mL crimp top vials with PTFE-coated septa
Aluminium seals
Sealcrimper and decapper
Tweezers
Analytical balance

2.2 Chemicals

Carbon disulfide, benzene-free
Dichloromethane for trace analysis
Methanol, analytical grade
The substances to be determined, analytical grade purity

2.3 Solutions

Desorption solution (ternary mixture: $CH_2Cl_2/CS_2/MeOH$ (60+35+5)): 60 parts dichloromethane, 35 parts carbon disulfide and 5 parts methanol are mixed in a suitably

sized volumetric flask. The mixture containing CS_2 should be stored in a cool place and in the dark to prevent yellowing or clouding of the desorption agent (beginning release of sulfur/colloidal sulfur from the CS_2).
The desorption solution can be kept for a maximum of one week.

2.4 Calibration standards

To calibrate the gas chromatographic system, five calibration standards are made from each substance to be analysed in concentrations at equidistant intervals within the linear operating range of the gas chromatographic system.
If the required concentration range (e.g. from one tenth to twice the threshold limit value) is greater than the linear operating range of the GC, the samples must be diluted accordingly.

Stock solution: First of all, from the components to be analysed a stock solution is prepared which contains 100 µL of the substance in 100 mL desorption solution. Determination of the amount of substance used should be carried out gravimetrically (using differential weighing). Alternatively, the weights of the substances can be calculated using the volumes added and the corresponding densities.
The procedure for the determination of a solvent mixture made up of ethylene glycol monomethyl ether, ethylene glycol monoethyl ether, toluene and ethylene glycol monobutyl ether is described below as an example.

Table 1. Pipetting scheme for the stock solution.

Substance	Added volume µL	Sample weight (theoretical) mg	Final volume mL	Theoretical concentration g/L
Ethylene glycol monomethyl ether	100	96.6	100	0.96
Ethylene glycol monoethyl ether	100	93.1	100	0.93
Toluene	100	87.0	100	0.87
Ethylene glycol monobutyl ether	100	90.0	100	0.90

Calibration standards: The calibration standards are prepared from the stock solution as shown in Table 2 (using ethylene glycol monomethyl ether as an example).

Table 2. Example of a pipetting scheme for the calibration standards for ethylene glycol monomethyl ether.

Calibration standard	Volume of the stock solution mL	Final volume mL	Concentration mg/L
1	0.1	100	0.96
2	0.2	20	9.6
3	0.2	10	19.2
4	0.5	10	48
5	1.0	10	96

Production of the calibration standards is also checked gravimetrically. In the solvent mixtures shown as examples, the following concentrations are obtained in the calibration standards for the substances to be determined:

Table 3. Concentrations of the calibration standards.

Calibration standard	Ethylene glycol monomethyl ether mg/L	Ethylene glycol monoethyl ether mg/L	Toluene mg/L	Ethylene glycol monobutyl ether mg/L
1	0.96	0.93	0.87	0.90
2	9.6	9.3	8.7	9.0
3	19.2	18.6	17.4	18.0
4	48	37.2	34.8	36.0
5	96	74.4	69.6	72.0

The stock solution is dispensed using a suitable syringe in an appropriate volumetric flask which already contains the desorption agent. The flask is then filled to the mark and the solution is mixed well.
The standards should be stored in the same way as the desorption mixture (refrigerator, in the dark). The shelf-life is about one week.

3 Sample collection and preparation

3.1 Sample collection

With a pump with adjustable flow rate or equipped with a gasmeter, air is drawn through an charcoal tube at a constant flow rate of 80 mL/min for a maximum sampling time of 2 h. During sampling, the charcoal tube is placed vertically in a suitable tube holder with the inlet at the top in the breathing zone of the person.

The parameters important for determining the concentration, such as sampled air volume, temperature and air pressure, must be determined. The sampling tubes are closed after sampling with the plastic caps provided and labelled clearly, and the sampling data and data on the work area/factory are noted in the sampling protocol. The output of the pump must also be checked and recorded after the end of sampling. It should not differ by more than 5% from the value originally set [1].
Loaded collection phases should not be stored for more than 14 days and should be kept in the dark. To prevent substance breakthrough, the sampling volume or the sample flow rate should not be greatly exceeded.

3.2 Preparation of the sample

To prepare the samples, the charcoal of the sampling systems is transferred as quickly and with as little loss as possible into a 5 mL crimp top vial and the vial is closed immediately. The vials should be stored in the refrigerator until analysis. Only after storage for four weeks are losses to be expected.
The charcoal with the sample is transferred to a crimp top vial, the vial is closed with a septum and 3 mL of the desorption mixture is added with a syringe. To determine the blank value an unloaded tube from the same batch as the sample tubes must be prepared. The charcoal tube is opened with a glass cutter, the contents are transferred to a 5 mL beaded rim vial, 3 mL of the desorption mixture is added and the vial is then closed.
The mixtures are left to stand for at least 30 minutes to desorb, during which period they are occasionally shaken. The supernatant is then decanted as soon as possible into two 1 mL crimp top vials and the vials are tightly closed.
The inclusion of a few charcoal particles does not usually interfere with the determination. The eluate is analysed using capillary GC/FID.
The desorbed samples should be analysed immediately.

4 Operating conditions for gas chromatography

Column:	Material:	fused silica, 0.31 mm internal diameter
	Length:	25 m
	Load:	1.03 µm HP-1 (methyl silicon)
Detector:		Flame ionisation detector (FID)
Carrier gas:	Helium:	50 kPa (1.3 mL/min)
Make-up gas:	Helium:	30 mL/min
Split:		20 mL/min
Temperatures:	Detector:	290 °C
	Oven:	10 min 35 °C isothermal, 5 °C/min up to 80 °C hold for 2 minutes, 10 °C/min up to 130 °C, 30 °C/min up to 240 °C hold for 2 minutes
	Injector:	230 °C
Injection volume:		3 µL

5 Analytical determination

The gas chromatograph is set up as given in the operating conditions in Section 4. With the autosampler (rinse: CH_2Cl_2) 3 µL of the sample is injected from the 1 mL vials into the gas chromatograph.
For routine purposes the samples only need be determined once. To check the calibration, at least one calibration solution per day should be regularly analysed along with the samples.

6 Calibration

Each calibration solution (see Section 2.3) is injected into the gas chromatograph twice and analysed like the sample solution. The injection volume is 3 µL. To draw the calibration curves, the peak areas are determined using an integrator system and plotted against the corresponding substance concentrations. The calibration curves are linear in the given concentration ranges for the substances tested. The calibration curves should be checked regularly, e. g. during routine analysis by checking at least one point. Blank values must be taken into consideration.

7 Calculation of the analytical result

7.1 Calculation of the analytical result after active sampling

From the peak areas obtained, the weights of the individual components are read from the appropriate calibration curves. The concentrations by weight ρ are calculated according to the following equation:

$$\rho = \frac{X}{V \cdot \eta} \cdot \frac{273 + t_g}{273 + t_a} \tag{1}$$

At 20 °C and 1013 hPa:

$$\rho_0 = \rho \cdot \frac{273 + t_a}{293} \cdot \frac{1013\ \text{hPa}}{p_a} \tag{2}$$

where:

ρ is the concentration by weight in mg/m^3
ρ_0 is the concentration by weight in mg/m^3 at 20 °C and 1013 hPa
X is the weight of the component in the solution in µg

V is the sampled air volume in L (determined with a gasmeter or calculated from the sampling time and flow rate)
η is the recovery
t_g is the temperature in the gasmeter in °C
t_a is the temperature of the ambient air in °C
p_a is the atmospheric pressure of the ambient air in hPa

8 Reliability of the method

8.1 Accuracy

Comparative experiments were carried out using solvent mixtures to test the accuracy of the method. For this purpose in a laboratory a dynamic test gas was produced by continuous injection [2].
The sampling tubes were each loaded for 2 hours with about 10 L of the test gas (40% relative humidity). After loading in parallel, the tubes were sent to a number of laboratories and analysed there according to the method described here. The results are shown in Table 4.

Table 4. Accuracy tests.

Substance	Load	Found			
		Examiner I		Examiner II	
	mg/m^3	mg/m^3	%	mg/m^3	%
Ethyl acetate	133.8	139.4	104.6	133.5	100.1
Isobutyl acetate	81.0	83.9	103.9	79.5	98.5
m-Xylene	32.1	32.6	101.7	31.5	98.3
2-Butanone	45.0	18.7	41.7	16.0	35.7
Hexone	29.8	29.3	98.5	27.5	92.5
Ethylene glycol monomethyl ether acetate	55.4	51.7	93.3		
Ethylene glycol monoethyl ether acetate	126.5	137.9	109.0		
Ethylene glycol monobutyl ether acetate	154.1	144.7	94.0		
Ethylene glycol monomethyl ether	29.3	24.9	84.9		
Ethylene glycol monoethyl ether	69.2	54.6	78.9		
Ethylene glycol monobutyl ether	80.5	73.3	91.0		

The method was not found to be wholly suitable for 2-butanone as significant substance losses occur when the sample air is moist and the samples are stored for more than one day.

8.2 Precision

In the comparative experiments described in Section 8.1 also the relative standard deviations (including sampling) were determined for 5 or 6 parallel samples.

Table 5. Standard deviation (relative) s, $n = 5$ or 6 determinations.

Substance	Load	Standard deviation (rel.) s	
		Examiner I	Examiner II
	mg/m^3	%	%
Ethyl acetate	133.8	1.0	3.3
Isobutyl acetate	81.0	1.1	3.7
m-Xylene	32.1	1.8	3.8
2-Butanone	45.0	6.5	8.4
Hexone	29.8	1.0	3.1
Ethylene glycol monomethyl ether acetate	55.4	4.5	
Ethylene glycol monoethyl ether acetate	126.5	3.9	
Ethylene glycol monobutyl ether acetate	154.1	3.0	
Ethylene glycol monomethyl ether	29.3	9.0	
Ethylene glycol monoethyl ether	69.2	3.6	
Ethylene glycol monobutyl ether	80.5	3.9	

8.3 Recovery rate

In comparative studies the percentage desorption of the substances investigated was determined with test gas.

Table 6. Desorption [%].

Substance	Load mg/m^3	Desorption %
Ethyl acetate	133.8	103.0
Isobutyl acetate	81.0	101.0
m-Xylene	32.2	101.0
2-Butanone	45.1	95.0
Hexone	29.8	101.0
Ethylene glycol monomethyl ether acetate	120.3	90.9
Ethylene glycol monoethyl ether acetate	274.8	107.1
Ethylene glycol monobutyl ether acetate	334.8	93.5
Ethylene glycol monomethyl ether	63.7	83.4
Ethylene glycol monoethyl ether	180.4	66.8
Ethylene glycol monobutyl ether	174.9	88.2

8.4 Quantification limit

For the glycol ethers quantification limits were determined of between 0.4 and 1.1 mg/m^3 for an air sample volume of 28 L.

Table 7. Calculated quantification limits.

Substance	Quantification limit mg/m^3
Ethyl acetate	0.8
Isobutyl acetate	0.5
m-Xylene	0.4
2-Butanone	0.6
Hexone	0.4
Ethylene glycol monoethyl ether acetate	1.1
Ethylene glycol monobutyl ether acetate	0.7
Ethylene glycol monomethyl ether	0.9
Ethylene glycol monoethyl ether	0.8
Ethylene glycol monobutyl ether	0.5

8.5 Specificity

As a result of the low specificity of the flame ionisation detector, interference from components with the same retention time can readily occur. To check the results, a second column with a different polarity is therefore used. If necessary a mass spectrometric detector could also be used.

9 Discussion of the method

The described procedures allow rapid, automatable and sufficiently exact determination of the components given in Table 7.
If relevant unknown components are present in a sample, these must be identified using GC/MS analysis.
The advantage of charcoal as adsorbent is that also apolar hydrocarbons are adsorbed. The use of the ternary desorption mixture allows a high desorption efficiency for a wide spectrum of polar and apolar solvents, i. e. the procedure is suitable for detecting a wide solvent spectrum in the ambient air. The procedure is also suitable for passive sampling.
The number of analyses that can be carried out each day is limited by the duration of a chromatographic run (cycle including cooling is about 45 minutes). Care must be taken that contamination of the samples, e. g. from contaminated ambient air while transferring the charcoal or from pens containing solvents while labelling the samples, is avoided.

10 References

[1] Europäisches Komitee für Normung (CEN): DIN EN 1232-Arbeitsplatzatmosphäre – Pumpen für die personenbezogene Probenahme von chemischen Stoffen – Anforderungen und Prüfverfahren. Brüssel 1993. Beuth Verlag, Berlin 1993.

[2] Verein Deutscher Ingenieure (VDI): VDI-Richtlinie 3490-Prüfgase, Blatt 1–16, Beuth Verlag, Berlin.

Authors: *J. Angerer, E. Flammenkamp, R. Hebisch, W. Krämer, M. Kuck, N. Lichtenstein, U. Risse, U. Schröter, M. Tschickardt*

Examiners: *E. Flammenkamp, K. Goßler, R. Hebisch, M. Hennig, U. Knecht, W. Krämer, M. Kuck, U. Schröter, M. Tschickardt*

Solvent mixtures

Method number	4
Application	Air analysis
Analytical principle	Gas chromatography (activated carbon/head-space)
Completed in	June 1997

Summary

The method described here is for the analysis of solvent mixtures in the air at the workplace by head-space gas chromatography [1] after adsorption on activated carbon. The gaseous solvent mixture is drawn through a sampling tube filled with activated carbon with a sampling pump and the components to be determined are deposited quantitatively on the adsorbent. After sampling, the components are desorbed by the addition of dimethylformamide, dimethylacetamide, benzyl alcohol or phthalic acid dimethyl ester and determined gas chromatographically with a flame ionisation detector. Quantitative evaluation is carried out using a calibration curve or by single-point calibration with an external calibration standard.

Precision of the whole procedure:	Standard deviation (rel.):	s = 0.8–6.2%
	Mean variation:	u = 2–15.5%
	in the concentration range from 0.1 to 2 times the MAK value (for the substances given in Section 8.2) and where n = 6 determinations	
Limit of quantification:	approx. 0.1 mg/m^3 (dependent on vapour pressure) with a sampled air volume of 120 L	
Recovery:	η > 0.85 (>85%) (for the substances given in Section 8.3)	
Sampling recommendation:	Sampling time:	2 hours
	Sample volume:	120 L
	Flow rate:	60 L/h

Solvent mixtures

See general section on methods 1–6 "Solvent mixtures, Introduction"

Authors: *J. Angerer, E. Flammenkamp, R. Hebisch, W. Krämer, M. Kuck, N. Lichtenstein, U. Risse, U. Schröter, M. Tschickardt*
Examiners: *E. Flammenkamp, K. Goßler, R. Hebisch, M. Hennig, U. Knecht, W. Krämer, M. Kuck, U. Schröter, M. Tschickardt*

Solvent mixtures

Method number 4

Application Air analysis

Analytical principle Gas chromatography (activated carbon/head-space)

Completed in June 1997

Contents

1 General principles

The gaseous solvent mixture is drawn into a tube filled with activated carbon with a sampling pump and the components to be determined are deposited quantitatively on the adsorbent. After sampling, the components are desorbed by the addition of dimethylformamide, dimethylacetamide, benzyl alcohol or phthalic acid dimethyl ester and determined gas chromatographically with a flame ionisation detector. Quantitative evaluation is carried out using a calibration curve or by single-point calibration with an external calibration standard. A desorption agent should be chosen whose substance peak (or relevant impurities) does not disturb the determination of the solvent components of interest.

2 Equipment, chemicals and solutions

2.1 Equipment

Gas chromatograph equipped with flame ionisation detector, possibly with mass selective detector and head-space autosampler, integrator/computer
Capillary columns with different polarity
Pump with flow rate adjustable to at least 0.5 to 2 L/min
Activated carbon adsorption tubes suitable for air sample flow rates up to 2 L/min (e.g. from SKC 226-36: length 150 mm, external diameter 8 mm, main zone 700 mg activated carbon, control zone 300 mg)
Gasmeter or stop clock and soap bubble flowmeter
Thermometer
Hygrometer
Barometer
Head-space vials with PTFE-coated septa
Aluminium crimp caps
Crimping tongs for sealing and opening the vials
Various volumetric flasks
Various pipettes
Glass cutter

2.2 Chemicals

The substances to be determined as shown in Table 2, analytical grade purity
Dimethylformamide, analytical grade or
Dimethylacetamide, analytical grade or
Benzyl alcohol, analytical grade or
Phthalic acid dimethyl ester, analytical grade

2.3 Calibration standard

Calibration standards which cover the concentration range of each of the constituents in the sample solution are necessary to calibrate the gas chromatographic determination. These are calculated from the sampled air volume and the volumes of the sample solution in accordance with the stipulations of the standard EN 482. The concentration range must cover at least the concentration range from 0.1 to 2 times the threshold limit value of the components in the air. To obtain a sufficiently high level of precision, the range of concentrations of the calibration standards should correspond to 0.01 times up to a maximum of 10 times the threshold limit values of the components to be determined.
In routine analysis it is usual to carry out quantitative determination using single-point calibration instead of calibration curves, as the drawing up of calibration curves is time-consuming and the calibration factors are subject to temporal changes depending on the apparatus used, whereas the course of the calibration curve (linearity) rarely changes. Care must be taken with single-point calibration that the concentrations for determination and calibration of the components are close enough to keep extrapolation errors as small as possible. This is guaranteed if the signals measured for the calibration standard and the sample solution do not differ by more than half a power of ten, i.e. by about a factor of 3 in both extrapolation directions.

2.3.1 Preparation of the calibration standards

Preparation of the calibration standards is described here using one component as an example. The procedure can be applied to several components if these can be determined together during one analytical run.

Calibration example:	
Substance:	Toluene
Density:	approx. 0.87 g/mL
Threshold limit value:	190 mg/m^3 (according MAK 2001)
Suggested volume of air sampled:	10 L
Necessary volume of desorption agent:	5 mL

Stock solution: A solution of about 0.2% of the liquid components in desorption agent is prepared. The following procedures are possible:

a) If the density of the components is known, 200 µL of the substance is injected with a microliter syringe below the surface of a desorption agent contained in a 100 mL calibrated volumetric flask, which is not filled to the calibration mark. The flask is then filled and the solution homogenised by shaking. The concentration of the stock solution is expressed in mg/L.

b) In a 10 mL vial closed with a septum into which 10 mL desorption agent has been added with a pipette and which has been weighed to exactly 0.1 mg, 20 µL of the substance is injected below the surface with a microliter syringe. The vial is shaken vigorously. Then the injected weight is determined to 0.1 mg exactly by differential weighing.

Solvent components have a high vapour pressure. For this reason weighing the sample directly with subsequent dilution is prone to error and should be avoided whenever possible!
The stock solution prepared according to either procedure (a) or (b) above contains the components with a concentration of about 1750 mg/L.
Concentration of the stock solution (c_0): approx. 1750 mg/mL

Calibration standard: In the following example, the calculation of the dilutions of the stock solution to produce the required calibration concentrations is described. The necessary calibration concentrations cover the whole concentration range to be analysed (see above). If necessary the sampling and preparation parameters must be adapted for the individual substance.

Necessary calibration standard concentrations:

for approx. 1 times the threshold limit value (c_1):	approx. 1.75 mg/5 mL and 350 mg/L
for approx. 0.1 times the threshold limit value (c_2):	approx. 0.175 mg/5 mL and 35 mg/L
for approx. 0.01 times the threshold limit value (c_3):	approx. 0.0175 mg/5 mL and 3.5 mg/L

Dilution of the stock solution: Calibration standard 1 for approx. 1 times the threshold limit value ($c_1 \cong 350$ mg/L): 20 mL stock solution is added to a 100 mL calibrated volumetric flask with a pipette, the flask is filled to the calibration mark with desorption agent and the solution is homogenised.

Calibration standard 2 for 0.1 times the threshold limit value ($c_2 \cong 35$ mg/L): About 50 mL desorption agent is placed in a 100 mL calibrated volumetric flask, 2 mL stock solution is added, the flask is filled to the calibration mark with desorption agent and the solution is homogenised.

Calibration standard 3 for 0.01 times the threshold limit value ($c_3 \cong 3.5$ mg/L): About 50 mL desorption agent is placed in a 100 mL calibrated volumetric flask, 200 µL stock solution is added, the flask is filled to the calibration mark with desorption agent and the solution is homogenised.

To prevent systematic dilution errors, it is strongly recommended that neither calibration concentration c_2 nor calibration concentration c_3 should be produced by repetitive diluting of calibration solution 1 (or calibration solution 2)!
With a maximum of three calibration standards of different concentrations, which correspond to 0.01 times, 0.1 times and 1 times the threshold limit value, the concentration ranges of, for example, 0.003 to 0.03 times the threshold limit value, of 0.03 to 0.3 times the threshold limit value and of 0.3 to 3 times the threshold limit value can be covered (i.e. 3 orders of magnitude) (see Tab. 1).

Table 1. Concentration of the calibration standards.

Calibration standard	Concentration	Corresponding to	Calibration range according
c_0	1750 mg/L	Stock solution	
c_1	175 mg/L	≈ MAK	0.3 to 3*MAK
c_2	17.5 mg/L	≈ 0.1*MAK	0.03 to 0.3*MAK
c_3	1.75 mg/L	≈ 0.01*MAK	0.003 to 0.03*MAK

3 Sample collection and preparations

3.1 Sample collection

With the flowmeter, the pump is set to obtain the necessary linear flow rate and the setting is recorded. An activated carbon tube is opened at both ends and connected to the pump in the direction stipulated. During sampling, the sampling tube is placed vertically in a suitable tube holder with the inlet at the top in the breathing zone of the person or in a static position.
Sampling should only be carried out at a relative humidity of <90% and an air temperature of <40 °C. The parameters important for determining the concentration, such as sampled air volume, temperature and air pressure, must be determined and recorded. The tubes are closed after sampling with the plastic caps provided and labelled clearly, and the sampling data and data on the work area/factory are noted in the sampling protocol. The analytical determination should be carried out within the tested shelf-life period of 7 days.

3.2 Preparation of the sample

The activated carbon phase with the sample is transferred to a head-space vial and 5 mL of the desorption mixture is added (see Section 2.2). The temperature in the sampling vessels is then allowed to come to equilibrium in the head-space gas chromatograph for 45 min at 90 °C.

4 Operating conditions for gas chromatography

The gas chromatographic operating conditions may have to be individually adjusted to suit the substances contained in the solvent mixture. Before injection from the head-space, the sample solution is equilibrated in a head-space vial for 45 minutes at 90 °C. The sample is then injected automatically from the vapour phase into the gas chromatograph.

Column:	50 m DB-5; internal diameter 0.3 mm; film thickness 0.1 μm	
Head-space parameters:	Temperature of the thermostat:	90 °C
	Equilibration time:	45 minutes
	Temperature of sampling needle:	130 °C
	Injection volume or time:	1 mL or 3 seconds
	Transfer line temperature:	150 °C
Carrier gas:	Nitrogen, column head pressure:	100 kPa
GC temperatures:	Injector:	150 °C
	Detector:	250 °C
	Column thermostat:	2 minutes at 50 °C, 10 °C/min up to 200 °C, 10 minutes at 200 °C

5 Analytical determination

The gas chromatograph and head-space autosampler are set up as described above and the samples and calibration standards (see Section 2.4) analysed under the same conditions. Evaluation of the peak areas of the solvent components to be determined is carried out using a suitable integration system.

6 Calibration

When preparing the calibration curve, each calibration standard (see Section 2.4) should be injected into the gas chromatograph three times. The injection volume and all set parameters correspond to those for the samples. For calculation, the mean value of the peak areas obtained is used.
The calibration curves are linear in the given concentration range for the components tested. In routine analysis the calibration curve must be checked regularly at least one point. Care must be taken, however, with single-point calibration that the concentrations of the components in the samples and in the reference calibration standard are close enough to keep extrapolation errors as small as possible.

7 Calculation of the analytical result

The weights of the individual components in the samples investigated are calculated from the peak areas obtained and the corresponding calibration curves. The concentrations by weight ρ are calculated according to the following equation:

$$\rho = \frac{X}{V \cdot \eta} \cdot \frac{273 + t_g}{273 + t_a} \tag{1}$$

At 20 °C and 1013 hPa:

$$\rho_0 = \rho \cdot \frac{273 + t_a}{293} \cdot \frac{1013 \text{ hPa}}{p_a} \tag{2}$$

where:

ρ concentration by weight in mg/m^3
ρ_0 concentration by weight in mg/m^3 at 20 °C and 1013 hPa
X weight of the component in the solution in µg
V sampled air volume (determined with a gasmeter or calculated from sampling time and flow rate) in L
η recovery
t_g temperature in the gasmeter in °C
t_a temperature of the ambient air in °C
p_a atmospheric pressure of the ambient air in hPa

8 Reliability of the method

Validation of the method was carried out with air concentrations adjusted equivalent to the MAK values valid in 1995. At present 2001 the following MAK values apply: cyclohexane: 700 mg/m^3; 1,1,1-trichloroethane: 1100 mg/m^3; ethyl acetate: 1500 mg/m^3; isobutyl acetate: 480 mg/m^3; 2-butanone: 600 mg/m^3; hexone: 83 mg/m^3 [2].

8.1 Accuracy

Collaborative studies were carried out using a solvent mixture to test the accuracy of the method. In a laboratory a dynamic test gas was produced for this purpose using continuous injection [3]. Activated carbon tubes were loaded for 2 hours with about 1 L/min (for *n*-hexane to 1,1,1-trichloroethane and 0.1 L/min for ethyl acetate to hexone) of the test gas at 40% relative humidity. After loading in parallel, the tubes were sent to a number of laboratories and analysed there according to this method. The desorption agents used were:

Examiner I: 10 mL dimethylacetamide for *n*-hexane to 1,1,1-trichloroethane and 3 mL dimethylacetamide for ethyl acetate to hexone
Examiner II: 10 mL dimethylacetamide for *n*-hexane to 1,1,1-trichloroethane and 10 mL benzyl alcohol for ethyl acetate to hexone

The results of the comparative experiment are shown in Table 2:

Table 2. Accuracy tests.

Substance	Load	Found			
		Examiner I		Examiner II	
	mg/m^3	mg/m^3	%	mg/m^3	%
n-Hexane	18.5	18.1	97.6	17.7	95.6
Cyclohexane	105	96.8	92.3	103	98.4
Toluene	31.7	28.7	90.5	28.9	91.1
Dichloromethane	34.8	36.7	106	28.1	80.8
1,1,1-Trichloroethane	107	103	96.4	106	99.2
Ethyl acetate	133	138	103	131	97.9
Isobutyl acetate	80.8	84.7	105	77.8	96.3
m-Xylene	32.1	32.3	101	30.7	95.7
2-Butanone	44.9	26.7	59.5	40.6	90.3
Hexone	29.8	30.4	102	28.6	96.1

With dichloromethane, in the comparative experiment substance was found also on the control phase (6–16%) by both examiners. Examiner I gave the sum of both zones, examiner II only the weight found in the main zone. Examiner I stored the samples in the refrigerator for one month before analysis.
The method was not found to be wholly suitable for 2-butanone as significant substance losses occur relatively quickly when the samples are stored. For the data given in the table above, the samples were analysed by examiner II within one week, by examiner I, however, only after three weeks.

8.2 Precision

In the comparative experiments described in Section 8.1 also the relative standard deviations from $n = 6$ parallel samples were determined, including sampling in each case. To determine the uncertainty associated with the analysis in the range in question, the relative standard deviation of the method in the low (approx. 0.1 times the MAK value), middle (approx. 1 times the MAK value) and upper (approx. 3 times the MAK value) concentration range was determined by one of each of the examiners according to EN 482 [4].
For the substances *n*-hexane to 1,1,1-trichloroethane 1 µL was loaded on the front zone of each of six adsorption tubes. Then 120 L laboratory air was drawn through each tube at a flow rate of 1 L/min, the tubes were each prepared with 10 mL dimethyl acetamide and analysed.
For the substances ethyl acetate to hexone each of 6 activated carbon tubes were loaded with about 10 L test gas with a relative humidity of about 45%, desorbed with 10 mL benzyl alcohol and analysed as described.

Table 3. Standard deviation (relative) s, $n = 6$ determinations.

Substance	Load	Standard deviation (rel) s	
		Examiner I	Examiner II
	mg/m³	%	%
n-Hexane	18.5	2.8	2.4
Cyclohexane	105	3.3	2.1
Toluene	31.7	3.5	1.8
Dichloromethane	34.8	3.4	4.1
1,1,1-Trichloroethane	107	3.4	1.9
Ethyl acetate	133	2.4	3.3
Isobutyl acetate	80.8	3.1	3.5
m-Xylene	32.1	5.2	3.7
2-Butanone	44.9	5.6	5.7
Hexone	29.8	2.6	3.7

Table 4. Standard deviation (rel.) s for different concentrations, $n = 6$ determinations.

Substance	MAK value mg/m³	Conc. mg/m³	s %	Conc. mg/m³	s %	Conc. mg/m³	s %
n-Hexane	180	1.5	1.5	15	2.2	49	2.5
Cyclohexane	1050	8.3	0.8	83	1.8	178	1.5
Toluene**	190	3	6.2	30	3.8	100	2.6
Dichloromethane	360	2.8	2.3	28	1.8	94	0.5
1,1,1-Trichloroethane	1080	8.7	1.2	87	2.0	289	1.0
Ethyl acetate	1400	180	1.2	1800	0.7	3600	2.0
Isobutyl acetate*	950	109	1.1	1090	0.9	2180	1.8
m-Xylene*	440	43	1.9	432	1.0	864	2.4
2-Butanone	590	61	1.1	605	1.0	120	1.9
Hexone	400	0	1.0	400	0.9	802	1.8

* MAK value applies for all isomers; ** Example of the calibration standard in Section 2.3

8.3 Recovery rate

During the collaborative study the percentage desorption of the substances *n*-hexane to 1,1,1-trichloroethane was determined by examiner I in the low, middle and upper concentration range. The desorption was determined by calculating the ratios between the solvent weights loaded on the sample tubes and the analytical results.

Table 5. Desorption [%] with different concentrations (examiner I).

Substance	Load	Desorption [%] at		
	mg/m^3	0.1 times MAK	1 times MAK	3 times MAK
n-Hexane	18.5	95.7	92.7	97.8
Cyclohexane	105	97.6	94.7	98.8
Toluene	31.7	86.8	86.6	93.1
Dichloromethane	34.8	97.9	91.9	96.4
1,1,1-Trichloroethane	107	97.4	94.2	98.8

For the substances ethyl acetate to MIBK the recoveries were determined using test gas at three different concentrations (for each of $n = 6$ samples).

Table 6. Desorption [%] with different concentrations (examiner II).

Substance	Load	Desorption [%] at		
	[mg/m^3]	0.1 times MAK	1 times MAK	3 times MAK
Ethyl acetate	133.5	98.6	99.7	100.0
Isobutyl acetate	80.8	99.7	100.3	101.0
m-Xylene	32.1	96.5	96.7	96.8
2-Butanone	44.9	98.9	100.6	101.8
Hexone	29.8	99.6	100.1	101.0

8.4 Quantification limit

Under the given analytical conditions the quantification limit is about 0.1 mg/m^3 for an air sample volume of 120 L.

8.5 Specificity

As a result of the low specificity of flame ionisation detectors, interference from components with the same retention time is possible with some gas chromatographic procedures. To check the analytical results, a second column of different polarity should therefore always be used, or possibly also a mass spectrometric detector.

9 Discussion of the method

The head-space GC method described permits rapid, automatic determination of the given solvent components. The choice of desorption agent is based on the purity available (the levels of highly volatile contaminants) and the position of the desorption agent peak in the chromatogram, as well as on the desorption efficiency for certain groups of substances. Dimethyl acetamide and dimethyl formamide have proved to be most suitable for a wide polarity range of components to be analysed. The desorption efficiency for other desorption agents must be determined individually for each substance. If relevant unknown substances are present in a sample, these must be identified before analysis using a mass selective detector.

10 References

[1] Applied Headspace Gas Chromatography, Verlag Heyden&Son GmbH, 1980. ISBN 0-85501-488-1.
[2] Deutsche Forschungsgemeinschaft. List of MAK and BAT Values 2001. Commission for the Investigation of Health Hazards of Chemical Compounds, Report No. 37. WILEY-VCH-Verlag, Weinheim, Germany
[3] Verein Deutscher Ingenieure (VDI): VDI-Richtlinie 3490-Prüfgase, Blatt 1–16, Beuth Verlag, Berlin.
[4] Europäisches Komitee für Normung (CEN): DIN EN 482-Arbeitsplatzatmosphäre-Allgemeine Anforderungen an Verfahren für Messung von chemischen Arbeitstoffen. Brüssel 1994. Beuth Verlag, Berlin 1994.

Authors: *J. Angerer, E. Flammenkamp, R. Hebisch, W. Krämer, M. Kuck, N. Lichtenstein, U. Risse, U. Schröter, M. Tschickardt*
Examiners: *E. Flammenkamp, K. Goßler, R. Hebisch, M. Hennig, U. Knecht, W. Krämer, M. Kuck, U. Schröter, M. Tschickardt*

Solvent mixtures

Method number	5
Application	Air analysis
Analytical principle	Gas chromatography (thermal desorption)
Completed in	June 1997

Summary

With this method solvent mixtures in the workplace air are analysed after thermal desorption. Active sampling is carried out. For some substances passive sampling is also possible. For active sampling, measured air volumes are drawn with a sampling pump through sampling tubes filled with adsorbent. The solvent constituents present in the air are adsorbed during this process. In passive sampling, enrichment of the substance takes place by diffusion. The adsorbed components are desorbed thermally and then analysed by a gas chromatograph equipped with a flame ionisation detector. Calibration standards of known composition are used for the quantitative evaluation. There is a linear relationship of the peak areas to the concentrations of the individual components of the solvent mixtures.

Precision of the whole procedure:	Standard deviation (rel.):	s = 0.5–8% (see Section 8.2)
	Mean variation:	u = 1.3–20%
	in the concentration range from 0.1 to 2 times the MAK value and where n = 6 determinations	
Limit of quantification:	2–5 mg/m^3, for a sampled air volume of 100 mL (for the substances given in Tab. 3 in the appendix)	
Recovery:	η > 0.99 (>99%) (see Tab. 13 in the appendix)	
Sampling recommendation:	Sampling time:	30–120 minutes
	Sample volume:	100–200 mL

Solvent mixtures

See general section on methods 1–6 "Solvent mixtures, Introduction"

Authors: *J. Angerer, E. Flammenkamp, R. Hebisch, W. Krämer, M. Kuck, N. Lichtenstein, U. Risse, U. Schröter, M. Tschickardt*

Examiners: *E. Flammenkamp, K. Goßler, R. Hebisch, M. Hennig, U. Knecht, W. Krämer, M. Kuck, U. Schröter, M. Tschickardt*

Solvent mixtures

Method number 5

Application Air analysis

Analytical principle Gas chromatography (thermal desorption)

Completed in June 1997

Contents

1 General principles

With this method solvent mixtures in the workplace air are analysed after thermal desorption. Active sampling is carried out. For some substances passive sampling is also possible. For active sampling, measured air volumes are drawn with a sampling pump through sampling tubes filled with adsorbent. The solvent constituents present in the air are adsorbed during this process. In passive sampling, enrichment of the substance takes place by diffusion. The adsorbed components are desorbed thermally and then analysed by a gas chromatograph equipped with a flame ionisation detector. Calibration standards of known composition are used for the quantitative evaluation.

2 Equipment and chemicals

2.1 Equipment

Adsorption tubes made of stainless steel, 6.3 mm x 90 mm, 5 mm internal diameter
Sampling pump, flow rate 1–4 mL/min (e.g. Air Check, from MTC, Müllheim, with tube rack for pressure-controlled sampling or S-205 from DEHA, Friolzheim)
Thermometer
Hygrometer
Barometer
Diffusion caps (passive sampling) (e.g. from Perkin-Elmer, order number 126433)
Gas chromatograph equipped with thermal desorber unit (e.g. ATD 400, from Perkin-Elmer) and flame ionisation detector
Computerised data collection and integration system
Apparatus for dynamic calibration or test gas apparatus
Caps (e.g. Swagelock® with PTFE seals, PTFE or aluminium)
Capillary columns of different polarity (e.g. DB-Wax 30 m, 0.5 µm film thickness, 0.23 mm internal diameter, from Promochem, Wesel and/or 50 m PVMS 5.0 µm film thickness, 0.32 mm internal diameter, from Perkin Elmer, Überlingen)
Gasmeter or stop watch and soap bubble flowmeter
Analytical balance
100 mL Volumetric flasks
1, 5 and 10 mL Bulb pipettes
10 µL Microliter syringe

2.2 Chemicals

The substances to be determined, analytical grade purity
Purified or synthetic air (free of hydrocarbons)
Hydrogen, purity 99.999%

Helium (carrier gas), purity 99.996%
Adsorber resin XAD-4 0.2–0.4 mm (pre-treatment: heat for 16 hours in a stream of nitrogen at 150 °C; turns yellow) or Chromosorb 106 (60–80 mesh)
Tenax-TA® 60–80 mesh

2.3 Pretreatment of the adsorption tubes

The adsorption tubes are packed with 220 to 230 mg Tenax-TA®, or about 450 mg XAD-4. For pretreatment 15 g of the XAD-4 adsorber resin is purified in a suitable stainless steel tube in a stream of nitrogen or helium at a flow rate of 100 mL/min at 150 °C for 16 h. This amount of adsorbent is sufficient to pack about 30 adsorption tubes. Tenax-TA® does not need pretreatment. The adsorbent is fixed between 2 stainless steel sieves; one of the sieves is placed exactly 15 mm from the end of the tube. This end can therefore also be used for passive sampling (see Fig. 1) [1].
Before use, the adsorption tubes are heated at 250 °C (Tenax-TA®), or 150 °C (XAD-4) in the thermal desorber for checking the blank values. For storage they are closed with Swagelok® or aluminium caps.

2.4 Solutions

Mixtures are produced by weighing the individual components to be analysed in relationship to their threshold limit value in air. These solutions are used either undiluted for producing test gases or added directly to the adsorption tubes after dilution in a suitable solvent. To determine the components listed below, a solution is produced gravimetrically. For direct dosing this solution must be diluted in a ratio of 1 : 100 in a solvent having a different retention time.

Table 1. Concentration of the solutions.

Components	Sample weight g/100 mL	Concentration of the dilution 1 : 100 µg/10 mL
Isobutyl acetate	20.8879	20.8879
Toluene	6.9279	6.9279
n-Butyl alcohol	8.1786	8.1786
2-Butanone	12.5538	12.5538
n-Heptane	29.1566	29.1566

2.5 Test gases

For thermal desorption preferably test gases are used for calibration. There are different methods for the generation of test gas mixtures [2]. One way of producing test gases is continuous injection (see Fig. 2) [3]. A calibration solution appropriate for the analysis (see Section 2.4) is injected continuously into a dynamic test gas apparatus at 120 µL per hour. To produce different concentrations, the dilution gas flow rate is set at 400, 1000 and 5000 mL per minute. The apparatus set-up is shown in Figure 2. The adsorption tubes can be loaded with defined volumes of the test gas and used for calibration.

3 Sample collection

Sample collection can be performed as static sampling or as personal air sampling. Suitable adsorbents must be selected before sampling (see Tab. 2 in the appendix) [4–7]. The adsorption tubes must be heated in the thermal desorber before sampling because interfering substances released from the sealing material of the PTFE caps can be adsorbed on the collection phase after longer periods of storage. The adsorption tubes are opened at the beginning of sampling. The parameters, which are important for the determination of the concentrations in air (sample air volume, temperature, atmospheric pressure and relative humidity), are noted in a sampling protocol.
Sampling is carried out in the breathing zone. The opening of the adsorption tube should not be obstructed. After sampling, the adsorption tubes are closed with PTFE caps. These samples should be analysed immediately. If the samples are stored for a longer period of time until evaluation, they must be sealed with suitable caps.

Note: If Swagelok® caps are used, tubes loaded with e.g. benzene, toluene and m-xylene can be stored for several months. A storage life of at least 5 months has been demonstrated for numerous compounds [6]. With other substances the storage life must be checked beforehand.

The method has been tested and found usable at relative humidities of 5–80%.

3.1 Active sampling

With a sampling pump the air to be analysed is drawn continuously through the adsorption tube at a flow rate of 1–4 mL/min. After sampling, the loaded adsorption tube is closed at both ends with caps.

3.2 Passive sampling

Before sampling, the cap at the end of the tube intended for passive sampling is replaced by a diffusion cap (see Fig. 3). Sampling is carried out in the breathing zone. The opening of the adsorption tube should not be obstructed. A sampling time of 4–8 hours is recommended. After sampling, the loaded adsorption tube is closed at both ends with caps. For the use of diffusion samplers and their limitations see [1] and [7].

4 Operating conditions for gas chromatography

Column:	Material:	fused silica
	Length:	30 m (column A) and 50 m (column B)
	Stationary phase:	*Column A*: DB-Wax, film thickness 0.5 µm, internal diameter 0.23 mm *Column B*: PVMS, film thickness 5.0 µm, internal diameter 0.32 mm
	Detector:	Flame ionisation detector (FID)
	Carrier gas:	Helium, column head pressure: 125 hPa
Temperatures:	Detector:	250 °C
	Furnace:	10 min 50 °C isothermal, 8 °C/min up to 120 °C, hold for 1.2 min, 12 °C/min up to 200 °C, hold for 10 min

The retention times obtained under the given conditions for column A are listed in Table 3 (see appendix), those for column B in Table 4 (see appendix). If substances are not separated under these conditions, different columns or a different temperature programme can be selected (see also Section 4 in method 1 "Solvent mixtures"). If a column with a non-polar stationary phase (column B, e. g. also used in parallel) is used the determination of gasoline-type hydrocarbon mixtures is possible.
Figure 4 shows an example of a chromatogram of a solvent mixture for column A.

5 Analytical determination

Thermal desorption tubes need not be prepared separately. Preparation takes place in the thermal desorber. The adsorption tubes are put into a compatible thermal desorber, heated and the desorbed components are transferred to a packed cooling trap with the carrier gas. When desorption is complete, the cold trap is heated and the substance mixture is transferred to the GC column.
The instrument settings for the thermal desorber (ATD 400, from Perkin-Elmer) are shown below:

Table 5. Instrument settings for the thermal desorber (ATD 400, Perkin-Elmer).

Adsorbent	Tenax-TA®	XAD-4
Desorption temperature	250 °C	150 °C
Desorption time	10 minutes	5 minutes
Transfer line	100 °C	100 °C
Cooling trap (adsorption)	–30 °C	–30 °C
Cooling trap (injection)	300 °C	300 °C

Table 5. (continued)

Adsorbent	Tenax-TA®	XAD-4
Weight of the adsorbent in the cooling trap	20 mg Tenax-TA®	20 mg Tenax-TA®
Carrier gas	Helium	Helium
Input split	41 mL/min	
Desorb flow	10 mL/min	
Output split	28 mL/min	

The instrument settings have to be changed if other types of thermal desorbers are used. After setting up the thermal desorber and the gas chromatograph (see Sections 4 and 5) the calibration standards and the samples are analysed.

6 Calibration

When using test gases, the concentrations of the substances should be between 0.1 and 2 times of their threshold limit value. The calibration solution produced as described in Section 2.4 is continuously injected into the dynamic test gas apparatus at 120 µL per hour. To produce different concentrations, the dilution gas flow rate is varied from 400, 1000 and 5000 mL per minute. The adsorption tubes are each loaded with a sampled air volume of 100 mL. The following test gas concentrations and calibration weights were obtained:

Table 6. Test gas concentrations and calibration weights.

Dilution stream	400 mL/min		1000 mL/min		5000 mL/min	
	Conc. mg/m^3	Weight µg	Conc. mg/m^3	Weight µg	Conc. mg/m^3	Weight µg
Isobutyl acetate	1044	104.4	417.8	41.8	83.6	8.4
Toluene	346.4	34.6	138.6	13.9	27.7	2.8
n-Butyl alcohol	408.9	40.9	163.6	16.4	32.7	3.3
2-Butanone	627.7	62.8	251.1	25.1	50.2	5,0
n-Heptane	1458	145.8	583.1	58.3	116.6	11.7

For calibration using direct injection of solutions onto the adsorption tube [6, 8], an adsorption tube is connected to the injector of a gas chromatograph, 1 µL to 50 µL aliquots of the diluted calibration solution (see Section 2.4) are injected and the substances are transferred to the tube at a flow rate of 50 mL per minute.

The calibration weights in Table 7 are obtained for 1, 5, 10 and 50 µL volumes.

Table 7. Calibration weights with calibration using direct injection of solutions.

Components	Calibration weight [µg] at 1 µL	at 5 µL	at 10 µL	at 50 µL
Isobutyl acetate	2.1	10.4	20.9	104.4
Toluene	0.7	3.5	6.9	34.6
n-Butyl alcohol	0.8	4.1	8.2	40.9
1-Butanone	1.3	6.3	12.6	62.8
n-Heptane	2.9	14.6	29.2	145.8

To draw the calibration curve, the peak areas determined using an integration system are plotted against the calibration weights used.
The calibration curves are linear in the range indicated. Calibration curves should be checked regularly in routine analysis.

7 Calculation of the analytical result

Using the peak areas obtained, the corresponding weight X in µg is read from each calibration curve. The corresponding concentration by weight ρ is calculated according to the following equation:

$$\rho = \frac{X}{V \cdot \eta} \tag{1}$$

At 20 °C and 1013 hPa:

$$\rho_0 = \rho \cdot \frac{273 + t_a}{293} \cdot \frac{1013}{p_a} \tag{2}$$

where:

- ρ is the concentration by weight of a component in mg/m^3
- ρ_0 is the concentration by weight in mg/m^3 at 20 °C and 1013 hPa
- X is the weight of the component in the sample in µg
- V is the sampled air volume in litres (calculated from the flow rate and the sampling time, for passive samples see below)
- η is the recovery (with calibration using test gas and complete desorption 1 can be used as the recovery value)
- t_a is the temperature during sampling in °C
- p_a is the atmospheric pressure during sampling in hPa

For passive sampling the following applies:

$$U_m = \frac{60 \cdot D_l A}{Z} \tag{3}$$

where:
U_m is the sampling rate (mL/min)
D_l is the diffusion coefficient in air (cm/s^2)
A is the surface of the sampler (cm^2) (type ATD: 0.196 cm^2)
Z is the diffusion distance (cm) (type ATD: 1.5 cm)

The sample volume V (in litres) can then be calculated as follows:

$$V = \frac{U_m \cdot \text{Sampling time}}{1000} \text{ [Litres]} \tag{4}$$

Further calculation is carried out in the same way as for active sampling.

8 Reliability of the method

Validation of the method was carried out with air concentrations equivalent to the MAK values valid in 1995. At present (2000) the following MAK values apply: cyclohexane: 700 mg/m^3; dichloromethane: carcinogen category 3A; 1,1,1–trichloroethane: 1100 mg/m^3 [9].

8.1 Accuracy

To test the accuracy of the method, Tenax-TA® tubes loaded with benzene, toluene and *m*-xylene and certified by the BCR were used. The gas mixture (synthetic air/BTX) necessary for loading the calibration standards was produced dynamically with a test gas apparatus using the continuous injection procedure in accordance with VDI 3490. To calibrate the procedure, 100 mL calibration gas was drawn through each Tenax-TA® tube with a piston pump and analysed as described above. The calibration standards (BCR reference material) were treated in the same way. Table 8 lists the values obtained for three BCR tubes.
In addition the accuracy of the method was tested in comparative experiments with selected solvent mixtures. For this purpose, test gases each containing 5 substances were produced dynamically in a laboratory and a series of Tenax-TA® tubes were each loaded with 200 mL of this test gas (40% relative humidity at 20 °C). The tubes were then closed with Swagelok® screw caps. Each of 6 of the tubes loaded in parallel and 3 blind samples were sent to a number of laboratories and analysed there according to this method. The results are shown in Table 9.

Table 8. Accuracy tests using BCR reference material.

Sample	Concentration [µg/tube] Benzene	Toluene	*m*-Xylene
BCR-45	1.082	1.107	0.978
BCR-46	1.087	1.116	1.030
BCR-58	1.106	1.178	1.047
Mean value	1.092	1.134	1.018
Certified value	1.053	1.125	1.043

Table 9. Accuracy tests.

Substance	Load	Found					
		Examiner I		Examiner II		Examiner III	
	µg/sample	µg/sample	%	µg/sample	%	µg/sample	%
n-Heptane	20.9	20.9	100	20.8	99.5	21.0	101
2-Butanone	8.99	8.94	99.4	8.94	99.4	8.93	99.3
n-Butyl alcohol	5.86	5.94	101	5.82	99.7	6.14	105
Toluene	4.96	4.91	99.0	4.97	100	4.95	99.7
Isobutyl acetate	15.0	14.8	99.1	14.8	98.7	15.3	102
n-Hexane	3.65	3.74	103	3.69	101		
Cyclohexane	20.7	18.7	90.4	20.1	97.1		
Toluene	6.26	6.40	102	6.27	100		
Dichloromethane	6.86	7.10	104	7.22	105		
1,1,1-Trichloro-ethane	21.2	20.5	96.9	20.6	97.1		

8.2 Precision

In the comparative experiments described in Section 8.1 also the relative standard deviations (including sampling) were determined for $n = 6$ parallel samples.

Table 10. Standard deviation (relative) s, $n = 6$ determinations.

Substance	Load	Standard deviation (rel.) s (%)		
	µg/sample	Examiner I	Examiner II	Examiner III
n-Heptane	20.88	0.43	0.79	2.9
2-Butanone	8.99	0.89	1.1	6.0
n-Butyl alcohol	5.86	2.1	1.5	5.7
Toluene	4.96	0.67	9.5	2.4
Isobutyl acetate	15.0	0.91	0.78	2.3

Table 10. (continued)

Substance	Load µg/sample	Standard deviation (rel.) *s* (%) Examiner I	Examiner II	Examiner III
n-Hexane	3.65	2.3	0.80	
Cyclohexane	20.7	2.4	2.2	
Toluene	6.26	2.0	3.1	
Dichloromethane	6.86	3.4	0.70	
1,1,1-Trichloro-ethane	21.2	2.2	1.9	

To determine the uncertainty associated with the analysis, the relative standard deviation of the method in the low (approx. 0.1 times the MAK value), middle (approx. 1 times the MAK value) and upper (approx. 2 times the MAK value) concentration range was determined by an examiner according to EN 482 [10]. For this purpose three test gases of different concentrations (see Table 11) were produced dynamically in a test gas apparatus and the test gases drawn through each of 6 Tenax-TA® tubes at a flow rate of 2–4 mL/min for 120 minutes.

Table 11. Standard deviation (rel.) *s* for different concentrations, $n = 6$ determinations.

Substance	MAK value mg/m³	Conc. mg/m³	%	Conc. mg/m³	%	Conc. mg/m³	%
n-Heptane		17.4	8.0	161	2.2	306	5.3
Cyclohexane	1050	115	7.5	–	–	–	–
Toluene	190	41.4	6.7	371	2.9	741	4.2
Dichloromethane	360	31.9	5.6	333	2.3	665	5.0
1,1,1-Trichloroethane	1080	103	6.7	1003	3.0	2001	4.0

As cyclohexane breakthrough can take place on Tenax-TA® when other substances are present and sample volumes are high, the standard deviation was not determined for this substance. Also with high concentrations of 1,1,1-trichloroethane, care must be taken that breakthrough does not occur when using Tenax-TA®. XAD-4 or Chromosorb 106 have better adsorption properties for these substances.

The characteristics of the method – determined at one of the laboratories which participated in the comparative experiments – are listed in Table 12 in the appendix. Adsorption tubes filled with Tenax–TA® were loaded in test gas atmospheres and thermally desorbed.

8.3 Recovery rate

In the comparative experiments, the percentage desorption of the individual substances was determined by twice heating various tubes. For the substances named, over 99% desorption was found at the given concentrations. Data for recovery with samples produced with test gas can also be taken from Table 12 in the appendix from the columns "Theoretical value" and "Mean".

8.4 Quantification limit

The quantification limits under the conditions given in Section 4 were between 2 and 5 mg/m^3 (see Tab. 12 in the appendix). If lower quantification limits are needed, the split ratio of the thermal desorber must be changed.

8.5 Specificity

As a result of the low specificity of flame ionisation detectors, interference from components with the same retention time is possible with some gas chromatographic procedures.
To check the analytical results, as described above, a second column of different polarity can be used as well, or perhaps also a mass spectrometric detector.

9 Discussion of the method

The adsorption tubes must be heated in the thermal desorber before sampling because interfering substances released from the sealing material of the caps can be adsorbed on the collection phase after longer periods of storage. The pump capacity and the total air flow rate must be set in such a way that about 2/3 of the breakthrough volume is not exceeded during sampling.
The procedure was tested with Tenax-TA® and XAD-4 tubes. A different adsorption material (e.g. Chromosorb 106) can, however, sometimes be used. XAD-4 and Chromosorb 106 are comparable in their adsorption behaviour. If other adsorbents or adsorption tubes of other sizes are used, the breakthrough and retention volumes and the analytical characteristics must be checked. The reader is referred to the relevant literature [4–7, 11].
In addition to active sampling, the adsorption tubes can also be used for passive sampling. It must be remembered, however, that the retention volume of the component to be determined is over 100 L/g of adsorbent at 20 °C. Table 13 in the appendix lists sampling rates which were determined in laboratory and field experiments. Reference [7] lists more sampling rates for numerous organic compounds. Given adequate adsorptive power of the adsorbent, passive sampling rates can also be calculated [12]. In practice,

tests with test gas have, however, revealed deviations from the expected theoretical values in individual cases.
Loaded sampling phases, which are closed with PTFE caps, should be analysed immediately. In the comparative experiments, samples closed with Swagelok® caps were stored for up to 12 days at room temperature without significant losses. In the literature [6] also storage life of 5 months and more are described.
The reader is referred to reference number [13] for the requirements of pumps for personal air sampling.

10 References

[1] *Blome H* (1988) Möglichkeiten und Grenzen der Verwendung von Diffusionssammlern zur Probenahme gas- und dampfförmiger Stoffe in der Luft in Arbeitsbereichen. Staub-Reinhaltung der Luft 48: 177–181

[2] *Greim H* (Ed.) (1994) Analytische Methoden zur Prüfung gesundheitsschädlicher Arbeitsstoffe, Luftanalysen. Band 1, Spezielle Vorbemerkungen-Herstellungsverfahren für Prüfgase, Kap. 5, VCH-Verlagsgesellschaft, Weinheim.

[3] *Verein Deutscher Ingenieure* (VDI): VDI-Richtlinie 3490-Prüfgase, Blatt 1–16, Beuth Verlag, Berlin.

[4] *Brown R H, Purnell C J* (1979) Collection and Analysis of Trace Organic Vapour Pollutants in Ambient Atmospheres. The Performance of a Tenax-GC Adsorbent Tube. Journal of Chromatography, 178: 79–90

[5] *Perkin Elmer GmbH* (1994) Thermal Desorption Applications No. 40, Monitoring Volatile Organics Compounds (VOCs) in Air Using Pumped, Solid Sorbent Tubes, Thermal Desorption & Gas Chromatography. Überlingen.

[6] *Internationale Organisation für Normung* (ISO): ISO DIN 9976-Determination of concentrations of volatile organic compounds in air-Pumped sorbent tube/thermal desorption/capillary gas chromatograpic method. 1994.

[7] *HSE/CAR Working Group 5* (1995) Diffusive Uptake Rates on the Perkin Elmer Sorbent Tubes. The Diffusive Monitor, 7: 9–11

[8] *Health and Safety Executive* (1983) Methods for the Determination of Hazardous Substances. Benzene in air. Laboratory method using porous polymer adsorbent tubes, thermal desorption and gas chromatography. MDHS 22, HSE, London.

[9] *DFG* (2000) MAK- und BAT-Werte-Liste 2000. Senatskommission zur Prüfung gesundheitsschädlicher Arbeitsstoffe, Mitteilung 36. Wiley-VCH Verlag, Weinheim.

[10] *Europäisches Komitee für Normung* (CEN): DIN EN 482-Arbeitsplatzatmosphäre-Allgemeine Anforderungen an Verfahren zur Messung von chemischen Arbeitstoffen. Brüssel 1994. Beuth Verlag, Berlin 1994.

[11] *Brown R H* (1996) What is the Best Sorbent for Pumped Sampling-Thermal Desorption of Volatile Organic Compounds. Experience with the EC Sorbents Project. Analyst, 121: 1171–1175.

[12] *Pannwitz K-H* (1983) Diffusionskoeffizienten, Drägerheft 327, S. 6–13, Fa. Dräger (Ed.), Lübeck.

[13] *Europäisches Komitee für Normung* (CEN): DIN EN 1232-Arbeitsplatzatmosphäre – Pumpen für die personenbezogene Probenahme von chemischen Stoffen – Anforderungen und Prüfverfahren. Brüssel 1997. Beuth Verlag, Berlin 1997.

Authors: *J. Angerer, E. Flammenkamp, R. Hebisch, W. Krämer, M. Kuck, N. Lichtenstein, U. Risse, U. Schröter, M. Tschickardt*
Examiners: *E. Flammenkamp, K. Goßler, R. Hebisch, M. Hennig, U. Knecht, W. Krämer, M. Kuck, U. Schröter, M. Tschickardt*

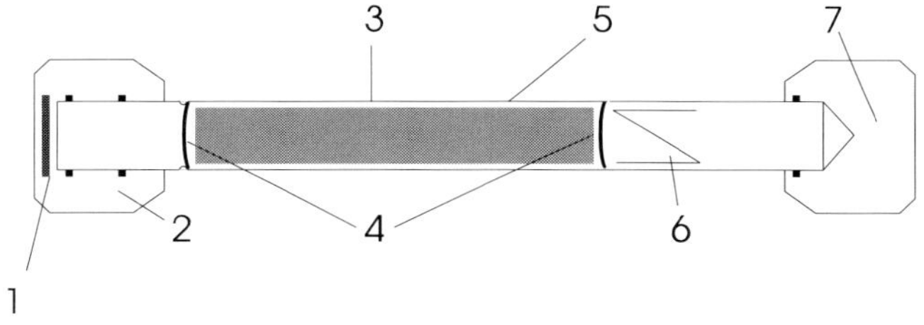

1 Sieve
2 Diffusion cap
3 Adsorbent
4 Stainless steel sieves
5 Adsorption tube
6 Spring
7 Cap

Fig. 1. Adsorption tube (type "ATD") with diffusion cap (passive sampling).

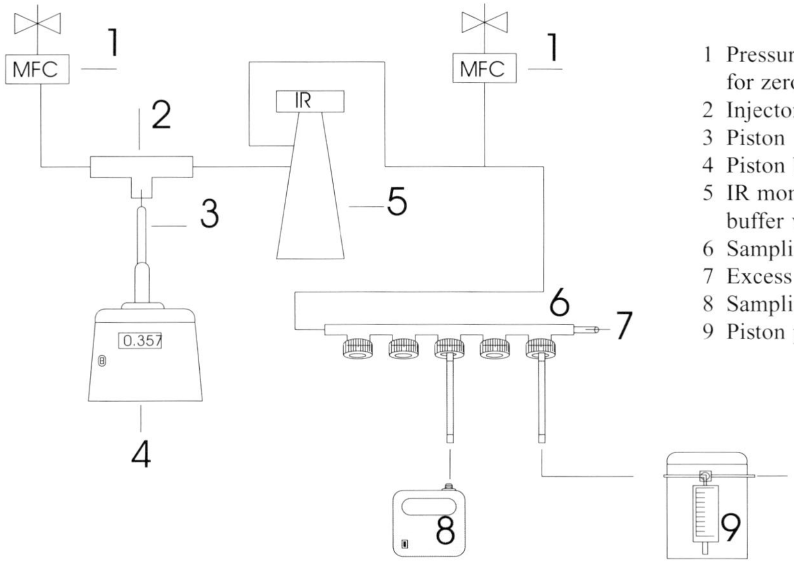

1 Pressure control for zero-gas
2 Injector
3 Piston
4 Piston burette
5 IR monitor or buffer vessel
6 Sampling manifold
7 Excess test gas
8 Sampling pump
9 Piston pump

Fig. 2. Test gas apparatus.

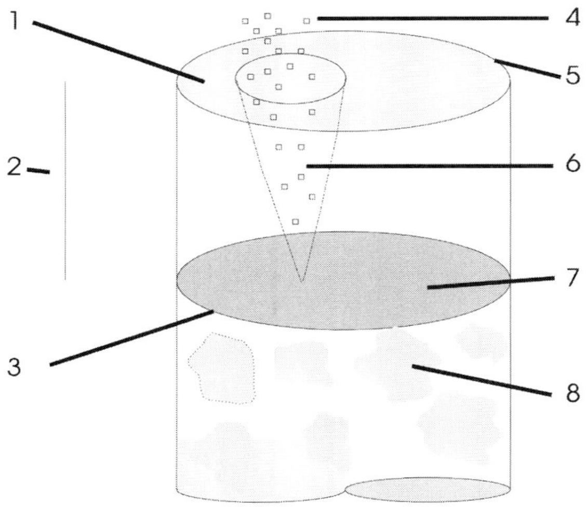

1 Surface of the sampler
2 Diffusion section
3 Concentration at the surface of the adsorbent
4 Concentration of the substance in the ambient air
5 Exposed end of the sampler
6 Concentration gradient
7 Sieve
8 Adsorbent

Fig. 3. Schematic diagram of a passive sampler.

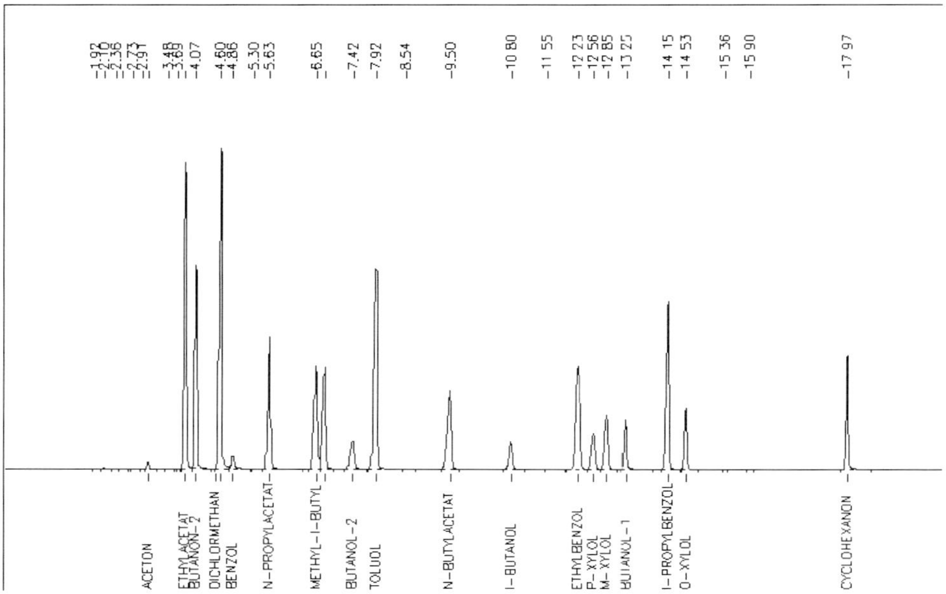

Fig. 4. Chromatogram for a solvent mixture (column A: DB-Wax, analytical conditions see Section 4).

Appendix

Table 2. Retention volumes and recommended maximum sampling volumes of various substances for the adsorbents Tenax-TA® and Chromosorb 106 at 20 °C [6].

	Tube filling: 200 mg Tenax-TA®		Tube filling: 300 mg Chromosorb 106	
Substance	Retention volume [L]	Maximum sampling volume [L]	Retention volume [L]	Maximum sampling volume [L]
n-Pentane			11.2	5.6
n-Hexane	6.4	3.2	60	30
n-Heptane	34	17	325	162.5
n-Octane	160	80	276	138
n-Nonane	1400	700	14000	7000
n-Decane	4200	2100	74000	37000
n-Undecane	25000	12500		
n-Dodecane	126000	63000		
Benzene	12.5	6.25	53	26.5
Toluene	76	38	165	82.5
Xylene	600	300	1554	777
Ethylbenzene	360	180	730	365
n-Propylbenzene	1700	850		
Isopropylbenzene	960	480		
Ethyltoluene	2000	1000		
Trimethylbenzene	3600	1800	5650	2825
Styrene	600	300		
Methyl styrene	2400	1200		
Carbon tetrachloride	12.4	6.2	44	22
1,2-Dichloroethane	10.8	5.4	34	17
1,1,1-Trichloroethane	not recommended		17	8.5
1,1,2-Trichloroethane	68	34		
Trichloroethylene	11.2	5.6		
Tetrachloroethylene	96	48		
Chlorobenzene	52	26		
Methyl acetate			5.2	2.6
Ethyl acetate	7.2	3.6	39	20
n-Propyl acetate	36	18	297	148.5
Isopropyl acetate	12	6	147	74
n-Butyl acetate	170	85	1460	730
Isobutyl acetate	265	132.5	880	440
tert-Butyl acetate	not recommended		327	164
Acrylic acid methyl ester	13	6.5		
Acrylic acid ethyl ester	48	24		
Methacrylic acid methyl ester	55	27.5		
Ethylene glycol monomethyl ether	6	3	9.6	4.8
Ethylene glycol monoethyl ether	10	5	150	75

Table 2. (continued)

Substance	Tube filling: 200 mg Tenax-TA®		Tube filling: 300 mg Chromosorb 106	
	Retention volume [L]	Maximum sampling volume [L]	Retention volume [L]	Maximum sampling volume [L]
Ethylene glycol monobutyl ether	70	35		
Propylene glycol 1-methyl ether	27	13.5		
Ethylene glycol monomethyl ether acetate	16	8	1720	860
Ethylene glycol monoethyl ether acetate	30	15	8100	4050
Ethylene glycol monobutyl ether acetate	300	150		
Acetone			2.4	1.2
2-Butanone	6.4	3.2	21	10
Hexone	53	26.5	490	250
Cyclohexanone	340	170		
Ethanol			2.4	1.2
n-Propyl alcohol			17	8
Isopropyl alcohol			9	4.5
n-Butyl alcohol	10	5	96	50
Isobutyl alcohol	5.6	2.8	60	30
tert-Butyl alcohol	not recommended			
n-Octanol	2800	1400		
Phenol	480	240		
Pyridine	16	8		
Aniline	440	220		
Nitrobenzene	28000	14000		

Note: Chromosorb and XAD-4 are comparable in their adsorption behaviour.

Table 3. Retention times of various solvents (column A: DB-WAX, analytical conditions see Section 4).

Time [min]	Substance
1.95	Trichlorofluoromethane (R-11)
2.00	1,1,2-Trichloro-1,2,2-trifluoroethane (R-113)
2.27	Ethylene oxide
2.52	Formic acid methyl ester
2.97	Acetone
3.09	Methyl acetate
3.53	Tetrahydrofuran
3.73	1,1,1-Trichloroethane
3.80	Isoflurane (1-chloro-2,2,2-trifluoroethyl difluoromethyl ether)
3.85	Ethyl acetate
3.91	Epichlorohydrin (1-chloro-2,3-epoxypropane)
4.02	Isopropyl acetate

Table 3. (continued)

Time [min]	Substance
4.08	2-Butanone (MEK)
4.15	Halothane (2-bromo-2-chloro-1,1,1-trifluoroethane)
4.55	Dichloromethane
4.65	Enflurane (2-chloro-1,1,2-trifluoroethyl difluoromethyl ether)
5.00	Benzene
5.84	*n*-Propyl acetate
6.36	Trichloroethylene
6.76	Methacrylic acid methyl ester
6.92	2-Hexanone
7.17	Isobutyl acetate
7.58	Tetrachloroethylene
7.73	*sec*-Butyl alcohol
8.38	Toluene
10.24	*n*-Butyl acetate
11.27	Isobutyl alcohol
12.80	Ethylbenzene
12.85	Propylene glycol 1-methyl ether
13.11	*p*-Xylene
13.37	*m*-Xylene
13.62	*n*-Butyl alcohol
14.61	Isopropyl benzene
14.75	Ethylene glycol monomethyl ether
14.97	*o*-Xylene
16.21	Ethylene glycol monoethyl ether
17.09	Ethylene glycol monomethyl ether acetate
17.21	Styrene
17.27	1,1,2-Trichloroethane
18.16	Ethylene glycol monoethyl ether acetate
18.18	Cyclohexanone
21.11	Ethylene glycol monobutyl ether
22.91	Ethylene glycol monobutyl ether acetate

Table 4. Retention times of various solvents (column B: PVMS, analytical conditions see Section 4).

Time [min]	Substance
3.34	Desflurane (2-difluoromethyl 1,2,2,2-tetrafluoroethyl ether)
4.64	Sevoflurane (1,1,1,3,3,3-hexafluoro-2-fluoromethoxypropane)
6.36	Isopropyl alcohol
8.11	Dichloromethane
11.66	*sec*-Butyl alcohol
11.83	2-Butanone
11.98	*n*-Hexane

Table 4. (continued)

Time [min]	Substance
12.86	Ethyl acetate
13.19	Chloroform
13.20	Isobutyl alcohol
14.91	1,1,1-Trichloroethane
15.34	*n*-Butyl alcohol
15.78	Benzene
16.37	Cyclohexane
16.85	Isooctane
17.38	*n*-Heptane
17.61	Trichloroethylene
17.76	*n*-Propyl acetate
18.88	Methylcyclohexane
19.14	2-Hexanone
20.57	Isobutyl acetate
20.84	3-Ethylhexane
20.87	Toluene
21.84	*n*-Octane
22.23	*n*-Butyl acetate
22.70	Tetrachloroethylene
24.28	Ethylbenzene
24.61	*m*-Xylene+*p*-Xylene
25.32	*n*-Nonane
25.41	*o*-Xylene
25.99	Isopropyl benzene
26.34	Cyclohexanone
27.79	*n*-Decane
30.45	*n*-Undecane
32.91	*n*-Dodecane

Table 12. Characteristic data of the method, determined by analysis of adsorption tubes loaded with test gas.

Substance	Concentration		Standard deviation (relative)	Quantification limit
	Theoretical value (load) mg/m^3	Mean (analysis) mg/m^3	s %	mg/m^3
Benzene	15	15	5.8	2
	2	2	2.4	
	1	0	not determined	
n-Butyl alcohol	78	77	5.7	4
	12	12	1.2	
	6	5	1.1	

Table 12. (continued)

Substance	Concentration		Standard deviation (relative)	Quantification limit
	Theoretical value (load) mg/m³	Mean (analysis) mg/m³	*s* %	mg/m³
sec-Butyl alcohol	91	89	5.8	4
	15	14	0.9	
	7	6	1.3	
Isobutyl alcohol	60	59	5.6	4
	10	9	1.4	
	4	4	10.9	
2-Butanone	369	364	6.3	4
	59	56	1.0	
	27	25	1.2	
Ethylene glycol monobutyl ether	248	256	3.0	4
	40	39	1.3	
	18	18	1.1	
Ethylene glycol monobutyl ether acetate	313	307	4.4	4
	50	57	2.7	
	23	34	1.5	
n-Butyl acetate	240	237	5.9	4
	38	36	1.8	
	18	16	1.6	
Isobutyl acetate	247	243	5.6	4
	40	37	1.6	
	18	17	1.4	
Cyclohexanone	149	148	5.0	4
	24	22	2.2	
	11	10	1.0	
Dichloromethane	49	49	0.9	5
Ethylene glycol monoethyl ether	191	200	3.5	4
	31	29	0.9	
	14	13	1.3	
Ethylene glycol monoethyl ether acetate	265	275	2.6	4
	41	41	1.0	
	19	18	1.1	
Ethyl acetate	820	807	6.3	4
	131	124	1.2	
	60	55	1.2	
Ethylbenzene	196	194	5.3	2
	31	30	1.8	
	14	13	0.8	

Table 12. (continued)

Substance	Concentration		Standard deviation (relative)	Quantification limit
	Theoretical value (load) mg/m³	Mean (analysis) mg/m³	*s* %	mg/m³
Isopropyl alcohol	812	800	6.3	4
	130	122	1.1	
	59	54	1.4	
n-Propyl acetate	309	305	6.1	4
	50	46	2.0	
	23	21	1.3	
Isopropyl acetate	850	847	3.9	4
	136	125	0.8	
	62	56	1.6	
Isopropyl benzene	202	199	5.0	2
	32	30	2.0	
	15	14	0.8	
Ethylene glycol monomethyl ether	51	54	3.8	
	4			
	8	8	3.2	
	4	4	1.8	
Ethylene glycol monomethyl ether acetate	68	70	3.1	4
	11	10	1.4	
	5	5	2.7	
Propylene glycol 1-methyl ether	838	871	3.3	4
	134	129	0.9	
	61	57	1.4	
2-Hexanone	236	233	5.7	4
	38	36	1.5	
	17	16	1.1	
Methacrylic acid methyl ester	489	510	3.5	4
	78	76	0.9	
	36	34	1.5	
Tetrachloroethylene	131	131	0.5	5
Tetrahydrofuran	1340	1389	3.5	4
	214	210	0.8	
	97	94	1.6	
Toluene	301	297	5.6	2
	48	46	1.5	
	22	20	0.5	
1,1,1-Trichloroethane	247	248	0.5	5
1,1,2-Trichloroethane	57	59	0.5	5

Table 12. (continued)

Substance	Concentration		Standard deviation (relative)	Quantification limit
	Theoretical value (load) mg/m^3	Mean (analysis) mg/m^3	s %	mg/m^3
Trichloroethylene	42	43	0.7	5
o-Xylene	67	66	5.1	2
	11	10	1.8	
	5	5	1.3	
m-Xylene	98	97	5.3	2
	16	15	1.7	
	7	7	0.5	
p-Xylene	65	64	5.3	2
	10	10	3.0	
	5	4	1.0	

Table 13. Passive sampling rates (for stainless-steel tube type "ATD" from Perkin Elmer).

Components	Adsorbent	Sampling rate mL/min
Tetrachloroethylene	XAD-4	0.4335 (over 7 days)
Styrene	Tenax-TA®	0.536

Table 14. Data for some adsorbents with thermal desorption. Retention volumes in litres per gram of adsorbent at 20 °C.

Substance	Breakthrough volumes* XAD-4 L/450 mg	Retention volumes* XAD-4 L/g	Tenax-TA® L/g
Benzene	31.6	267	61
Acetone	9.3	60	5.4
R-11	1.8	10.6	0.3
Methyl acetate	6.6	60	5.4
Dichloromethane	2.4	19	4
Methanol	1.7	13	0.3
Enflurane	27	238	not determined
Isoflurane	20	212	not determined
Halothane	16	169	1.5
Ethylene oxide	0.7	7.5	not determined
R-113	8.0	106	

* Breakthrough volume in litres per 450 mg XAD-4 at 30 °C

Solvent mixtures

Method number	6
Application	Air analysis
Analytical principle	Gas chromatography (head-space/silica gel)
Completed in	June 1997

Summary

The method permits the determination of volatile alcohols and 2-butanone in the air at the workplace by adsorption on silica gel and analysis by head-space gas chromatography. Desorption of the analytes from the silica gel is carried out with deionised water.

Precision of the whole procedure:	Standard deviation (rel.):	s = 0.8–3.3%
	Mean variation:	u = 2.0–8.3%
	in the concentration range from 0.1 to 2 times the MAK value (for the substances given in Section 8.2) and where n = 6 determinations	
Limit of quantification:	2–4% (methanol 6%) of the MAK values valid in 1995	
Recovery:	η > 0.90 (>90%)	
Sampling recommendation:	Sampling time:	0.5 h (–2 h)
	Sample volume:	6 L

Solvent mixtures

See general section on methods 1–6 “Solvent mixtures, Introduction“

Authors: *J. Angerer, E. Flammenkamp, R. Hebisch, W. Krämer, M. Kuck, N. Lichtenstein, U. Risse, U. Schröter, M. Tschickardt*

Examiners: *E. Flammenkamp, K. Goßler, R. Hebisch, M. Hennig, U. Knecht, W. Krämer, M. Kuck, U. Schröter, M. Tschickardt*

Solvent mixtures

Method number 6

Application Air analysis

Analytical principle Gas chromatography (head-space/silica gel)

Completed in June 1997

Contents

1 General principles

The alcohol or 2-butanone vapour in the air to be analysed is adsorbed on silica gel in adsorption tubes by means of a sampling pump equipped with a flow regulator. For the gas chromatographic determination the adsorbed substances are desorbed with deionised water. The analytical determination is carried out by head-space gas chromatography. Quantitative evaluation is carried out with a calibration curve.

2 Equipment, chemicals and solutions

2.1 Equipment

Head-space gas chromatograph equipped with flame ionisation detector (HS 100, from Perkin-Elmer, Überlingen)
Integrator/computer
Sampling pump (0.2 L/min or 50 mL/min)
Silica gel adsorption tubes (e.g. type G or type B from Dräger)
Stop clock and soap bubble flowmeter
Thermometer
Hygrometer
Barometer
Head-space vials with PTFE-coated septa
Aluminium crimp caps
Crimping tongs for sealing and opening the vials
10 mL Volumetric flasks
Various bulb pipettes
Various microliter syringes
Glass cutter
Shaking machine

2.2 Chemicals

The substances to be determined, analytical grade purity
Deionised water, free of interfering impurities

2.3 Calibration standards

Gas chromatographic determination is carried out by means of a calibration curve. Three to five calibration standards which cover the linear operating range of the gas chromatographic system are produced from each of the substances to be analysed.

If the required concentration range (e. g. from one tenth to twice the threshold limit value) is greater than the linear operating range of the gas chromatograph, the samples must be diluted accordingly. With a greater linear operating range, calibration is only carried out in the concentration range of interest.
Example of the preparation of calibration standards:

Stock solution:
First a stock solution is prepared by adding the following weights of substance to a 10 mL volumetric flask:
1 mL methanol
3 mL isopropyl alcohol
1 mL 2-butanone
Then the flask is filled to the mark with ethanol.

Calibration standards:
The calibration standards are produced by diluting the solution with deionised water. The concentrations in Table 1, for example, are then obtained.

Tab. 1. Concentration of the calibration standards.

	Sample weight	Dilution [µg/mL]			
	[mg/mL]	1:200	1:500	1:1000	1:2000
Methanol	80.59	402.95	161.18	80.59	40.30
Isopropyl alcohol	233.91	1962.35	784.94	392.49	196.24
2-Butanone	81.02	1169.55	467.82	233.91	116.96
Ethanol	392.49	405.10	162.04	81.02	40.51

Silica gel is added to each head-space vial from an unloaded tube (collection phase). Then 5 mL of the aqueous standard solution is added, the vials are immediately closed and – like the samples – equilibrated for 2 hours on the shaker; the silica gel in the solution should move slightly.

3 Sample collection and preparation

3.1 Sample collection

With a pump equipped with a flow regulator, air is drawn through silica gel adsorption tubes at a constant flow rate of 12 L/h. The long zone of the tube is taken as the collection phase.
A sampling time of 30 minutes should not be greatly exceeded as breakthrough can occur with higher concentrations. If longer sampling times are necessary, the flow rate must be decreased accordingly; for two hours sampling, for example, to 3 L/h.

During sampling, the adsorption tube is placed in a suitable tube holder in the breathing zone of the person or in a static position. The parameters important for determining the concentration, such as sampled air volume, temperature and air pressure, must be determined and recorded. When sampling is complete, the tubes are closed with the plastic caps provided and labelled clearly, and the sampling data are noted in the sampling protocol.

3.2 Sample preparation

Desorption of the silica gel adsorption tubes is carried out separately for the collection phase and control phase. The loaded silica gel is transferred directly to the head-space vial, 5 mL deionised water is added and the vial is immediately closed. The samples are then equilibrated for 2 hours at room temperature on the shaker.

4 Operating conditions for gas chromatography

For static head-space chromatography, the sample solution is equilibrated in the head-space vial for 12 minutes at 60 °C. Injection takes place from the vapour phase.

GC conditions:

Column:	50 m SE 54, 0.25 mm i.d., 0.5 μm film thickness
Carrier gas:	Nitrogen
Column head pressure:	50 kPa
Pressure in sample vial:	100 kPa
Temperature programme:	15 min. at 60 °C, with 30 °C/min. to 180 °C, 1 min. at 180 °C
Detector:	FID at 230 °C

Parameters for the head-space sampler (HS 100, Perkin-Elmer):

Thermostat temperature:	60 °C
Transfer line temperature:	90 °C
Thermostatting time:	12 minutes
Injection time:	0.08 minutes
Pressure time*:	0.5 minutes

* (Duration of the pressure build-up in the sample vial)

5 Analytical determination

The operating conditions for the head-space sampler and gas chromatograph are set up as given in Section 4. Calibration standards are determined in regular alternation with the samples.

Injection can only be carried out once from each vial. The collection phase and control phase are analysed separately according to the same scheme. If 10% or more of the analysed substance is found on the control phase, sampling must be repeated.
Evaluation of the peak areas is carried out with a suitable integration system.

6 Calibration

Calibration is carried out by means of an external standard.
At least three calibration solutions are analysed to draw up the calibration curve for each dilution (see Section 2.3). The calibration curves are drawn up from the amounts of the substance analysed and the mean values of the peak area values obtained.
The calibration curves are linear in the given concentration ranges for the substances tested. Calibration curves should be checked regularly, e. g. during routine analysis by checking at least one point. Blank values must be taken into consideration.

7 Calculation of the analytical result

From the weights of the individual components obtained, the concentration by weight ρ (mg/m^3) is calculated according to the following equation:

$$\rho = \frac{X}{V \cdot \eta} \cdot \frac{273 + t_g}{273 + t_a} \tag{1}$$

At 20 °C and 1013 hPa:

$$\rho_0 = \rho \cdot \frac{273 + t_a}{293} \cdot \frac{1013 \text{ hPa}}{p_a} \tag{2}$$

where:

ρ is the concentration by weight in mg/m^3
ρ_0 is the concentration by weight in mg/m^3 at 20 °C and 1013 hPa
X is the weight of the component in the solution in µg
V is the sampled air volume (determined with a gasmeter or calculated from the sampling time and flow rate) in L
η is the recovery
t_g is the temperature in the gasmeter in °C
t_a is the temperature of the ambient air in °C
p_a is the atmospheric pressure of the ambient air in hPa

8 Reliability of the method

Validation of the method was carried out with air concentrations equivalent to the MAK values valid in 1995. At present 2001 the following MAK values apply: methanol: 270 mg/m^3; isopropyl alcohol: 500 mg/m^3; *n*-butyl alcohol 310 mg/m^3; 2-butanone: 600 mg/m^3 [1].

8.1 Accuracy

The accuracy of the method was checked in comparative experiments with solvent mixtures. For this purpose in a laboratory a dynamic test gas was produced by means of continuous injection. Sampling tubes were loaded with test gas (40 % relative humidity) in parallel for 30 minutes at a flow rate of 200 mL/min. The tubes were sent to a number of laboratories and analysed there according to the method described here. The results are shown in Table 2.

Table 2. Accuracy tests.

Substance	Load	Found			
		Examiner I		Examiner II	
	mg/m^3	mg/m^3	%	mg/m^3	%
Methanol	38.1	35.4	92.9	34.3	90.0
Ethanol	187	171	91.3	174	93.1
Isopropyl alcohol	75.7	68.3	90.2	73.3	96.8
n-Butyl alcohol	38.2	34.4	90.0	37.0	96.8
2-Butanone	38.1	33.2	87.1	37.5	98.4

8.2 Precision

In the comparative experiments described in Section 8.1 also the relative standard deviations (Table 3) were determined for $n = 6$ parallel samples (including sampling).
To determine the uncertainty associated with the analysis in the range in question, the relative standard deviation of the method in the low (approx. 0.1 times the MAK value), middle (approx. 1 times the MAK value) and upper (approx. 2 times the MAK value) concentration range were determined by an examiner according to EN 482 [2] by loading each of 6 sampling tubes (Table 4).

Table 3. Standard deviation (relative) s, $n = 6$ determinations.

Substance	Load	Standard deviation (rel.) s	
		Examiner I	Examiner II
	mg/m³	%	%
Methanol	38.1	3.2	2.8
Ethanol	187.3	2.3	3.7
Isopropyl alcohol	75.7	2.8	3.8
n-Butyl alcohol	38.2	3.7	4.2
2-Butanone	38.1	3.8	3.4

Table 4. Standard deviation (rel.) s for different concentrations, $n = 6$ determinations.

Substance	MAK value mg/m³	Load mg/m³	s %	Load mg/m³	s %	Load mg/m³	s %
Methanol	260	27.6	3.2	266	0.8	534	1.0
Ethanol	1900	201	1.8	1880	1.0	3760	0.9
Isopropyl alcohol	980	108	2.2	1010	0.9	2030	0.9
n-Butyl alcohol	300	32.3	2.1	303	2.0	623	0.6
2-Butanone	590	47.8	3.3	595	0.9	1240	0.8

8.3 Recovery rate

During validation of the method, the recovery η of the individual substances in the low, middle and upper concentration ranges was determined (Table 5).

Table 5. Recovery [%] with different concentrations.

Substance	0.1 × MAK		1 × MAK		2 × MAK	
	Load mg/m³	η %	Load mg/m³	η %	Load mg/m³	η %
Methanol	28.5	97	285	93	570	94
Ethanol	199	101	1990	94	3970	95
Isopropyl alcohol	107	100	1070	94	2140	95
n-Butyl alcohol	33.0	98	330	92	660	95
2-Butanone	64.8	74	648	92	1300	96

8.4 Quantification limit

When the given sampling conditions are observed and with a sampled air volume of 6 L, 2–4% of the corresponding threshold limit values (1995) (methanol 6%) can be determined accurately.

8.5 Specificity

As a result of the low specificity of the flame ionisation detector, interference from components of the same retention time can occur. To check the results, a second column with a different polarity should be used.

8.6 Shelf-life

The loaded sampling tubes can be stored for 2 weeks.

9 Discussion of the method

The procedure described is easy to carry out and allows the automatic and exact determination of the lower molecular weight alcohols and ketones.
Interference can occur when moderately volatile substances (e.g. hydrocarbon mixtures) are present. In such cases the column should be cleaned by heating after each sample analysis. If necessary the final temperature of the temperature programme must then be set higher than 180 °C. Interference from water vapour has not been observed.

10 References

[1] Deutsche Forschungsgemeinschaft. List of MAK and BAT Values 2001. Commission for the Investigation of Health Hazards of Chemical Compounds, Report No. 37. WILEY-VCH-Verlag, Weinheim
[2] Europäisches Komitee für Normung (CEN): DIN EN 482-Arbeitsplatzatmosphäre-Allgemeine Anforderungen an Verfahren für Messung von chemischen Arbeitsstoffen. Brüssel 1994. Beuth Verlag, Berlin 1994

Authors: *J. Angerer, E. Flammenkamp, R. Hebisch, W. Krämer, M. Kuck, N. Lichtenstein, U. Risse, U. Schröter, M. Tschickardt*
Examiners: *E. Flammenkamp, K. Goßler, R. Hebisch, M. Hennig, U. Knecht, W. Krämer, M. Kuck, U. Schröter, M. Tschickardt*

Volatile inorganic acids (HCl, HBr, HNO_3)

Method number 1

Application Air analysis

Analytical principle Ion chromatography

Completed in April 1997

Summary

The method described by NIOSH [1] was modified to make it possible to determine the gaseous inorganic acids, hydrogen chloride, hydrogen bromide and nitric acid, in the concentration range from 0.1 to 3 times the currently valid threshold limit values in air [2,3].
During sampling, the air containing the acids is drawn through an adsorption tube filled with purified silica gel with a Teflon filter for collecting chloride, bromide and nitrate particles at the inlet. Elution is carried out with an aqueous sodium bicarbonate/sodium hydrogen carbonate solution. Quantitative determination is carried out by ion chromatography.

Quantification limit:	HCl absolute 15 ng = 0.1 mg/m^3 HBr absolute 15 ng = 0.1 mg/m^3 HNO_3 absolute 15 ng = 0.1 mg/m^3 for a sampled air volume of 30 L.
Recovery:	HCl, HBr, HNO_3 $\eta = 0.96–0.97$ (96–97%) for the whole concentration range.

Precision:

Tab. D 1. Standard deviation (rel.) *s* and mean variation *u*, *n* = 10 determinations.

Substance	Concentration mg/m³	Standard deviation (rel.) *s* %	Mean variation *u* %
HCl	1.5	7.3	16.3
	3.5	2.0	4.5
	7.0	7.3	16.2
HBr	3.5	2.4	5.3
	10.0	2.5	5.5
	20.0	3.6	8.0
HNO_3	0.5	2.7	5.9
	2.5	3.0	6.6
	5.0	3.6	8.1

Sampling recommendation: Sampling time: 2 hours
Sampled air volume: 30 L (at concentrations up to 3 times the threshold limit value in air)

Hydrogen chloride (HCl) [CAS No. 7647-01-0]

Hydrogen chloride is a colourless, highly caustic gas with a pungent smell (molecular weight 36.46, freezing point –114 °C, boiling point –85 °C). Hydrogen chloride dissolves in water with the generation of heat and forms the also strongly caustic hydrochloric acid.
Hydrogen chloride and hydrochloric acid are widely used in industry (e. g. for extraction of rock phosphates and other ores, in metal processing for pickling, etching, and soldering, in electroplating).
The currently valid MAK value is 5 mL/m^3 or 7.6 mg/m^3. Hydrogen chloride is classified in Peak limitation category I in the List of MAK and BAT Values and given an excursion factor of =1= according to TRGS 900 [2, 3].

Hydrogen bromide (HBr) [CAS No. 10035-10-6]

Hydrogen bromide is a colourless, non-inflammable, highly corrosive gas with a pungent smell (molecular weight 80.92, freezing point –87 °C, boiling point –67 °C). It is readily soluble in water and forms the highly caustic hydrobromic acid.

Hydrogen bromide is used in the production of inorganic and organic bromides, as reducing agent and catalyst.
The currently valid MAK value is 2 mL/m^3 or 6.7 mg/m^3. Hydrogen bromide is classified in Peak limitation category I in the List of MAK and BAT Values and given an excursion factor of =1= according to TRGS 900 [2,3].

Nitric acid (HNO_3) [CAS No. 7697-37-2]

Anhydrous nitric acid is a colourless liquid which fumes in moist air (molecular weight 63.02, freezing point −41.6 °C, boiling point 84.1 °C). Nitric acid is used in various concentrations mixed with water. Fuming nitric acid (>70%) is a yellow to reddish-brown, highly caustic liquid. Concentrated nitric acid (69.2%) is formed during the distillation of nitric acid as an azeotropic mixture with water. Nitric acids with concentrations <70% are water-white, highly caustic liquids with a pungent smell.
Nitric acid is widely used in industry (e.g. as an oxidising agent, in the production of fertilisers, in the nitration of organic chemicals).
The currently valid MAK value is 2 mL/m^3 or 5.2 mg/m^3. Nitric acid is classified in Peak limitation category I in the List of MAK and BAT Values and given an excursion factor of =1= according to TRGS 900 [2, 3].

Author: *D. Breuer*
Examiner: *R. Hebisch, K. Macho*

Volatile inorganic acids (HCl, HBr, HNO_3)

Method number	1
Application	Air analysis
Analytical principle	Ion chromatography
Completed in	April 1997

Contents

1 General principles

The method described by NIOSH [1] was modified to make it possible to determine the gaseous inorganic acids, hydrogen chloride, hydrogen bromide and nitric acid, in the concentration range from 0.1 to 3 times the currently valid threshold limit values in air [2, 3].
During sampling, the air containing the acids is drawn through an adsorption tube filled with purified silica gel with a Teflon filter for collecting chloride, bromide and nitrate particles at the inlet. Elution is carried out with an aqueous sodium bicarbonate/sodium hydrogen carbonate solution. Quantitative determination is carried out by ion chromatography.

2 Equipment, chemicals and solutions

2.1 Equipment

Pump for personal sampling, flow rate 15 L/h
Gasmeter
Ion chromatograph with suppressor and conductivity detector
Data processing unit
Syringes for injection of samples or autosampler
10 to 1000 mL Volumetric flasks
10 to 2500 μL Adjustable pipettors
20 mL Screw-cap vessels made of polyethylene
Disposable filters, pore size 0.45 μm
2 to 10 mL disposable syringes with needles of 60 × 0.6 mm
Ultrasonic bath
Ultrapure water system
Teflon filter, diameter 37 mm, pore size 0.45 μm (e.g. from Macherey-Nagel)
Silica gel tubes (e.g. ORBO 53, from Supelco)

2.2 Chemicals

Chloride standard solution β (Cl^-) = 1.000 g/L (e.g. from Merck, Order No. 19897)
Bromide standard solution β (Br^-) = 1.000 g/L (e.g. from Merck, Order No. 19896)
Nitrate standard solution β (NO_3^-) = 1.000 g/L (e.g. from Merck, Order No. 19811)
Sodium bicarbonate anhydrous, analytical grade, (e.g. from Fluka, Order No. 71350)
Sodium hydrogen carbonate, analytical grade, (e.g. from Fluka, Order No. 71627)

2.3 Solutions

The following solutions are prepared using ultrapure water (resistivity >17 MΩ · cm).

Stock solution of the eluent:
2.86 g Na_2CO_3 and 0.25 g $NaHCO_3$ are dissolved in water in a 100 mL volumetric flask and the flask is filled to the mark.

Eluent and elution solution:
10 mL of the stock solution is placed in a 1000 mL volumetric flask and the flask is filled to the mark with ultrapure water (0.0027 M Na_2CO_3, 0.0003 M $NaHCO_3$).

2.4 Calibration standards

Combined standard:
1 mL of each of the chloride, bromide and sodium standard solutions (β = 1 g/L) are placed in a 10 mL volumetric flask and the flask is filled to the mark with ultrapure water. The combined standard solution thus produced has a concentration of 100 mg/L of each anion.

Calibration solutions:
From the combined standard solution, the calibration solutions in Table 1 are produced by dilution.

Table 1. Calibration solutions.

Calibration solution No.	Volume of the combined standard solution µL	Final volume of the calibration solution mL	Concentration [mg/mL]		
			Chloride	Bromide	Nitrate
1	200	50	0.4	0.4	0.4
2	400	50	0.8	0.8	0.8
3	600	50	1.2	1.2	1.2
4	800	50	1.6	1.6	1.6
5	1000	50	2.0	2.0	2.0
6	1200	50	2.4	2.4	2.4
7	1400	50	2.8	2.8	2.8
8	1600	50	3.2	3.2	3.2
9	1800	50	3.6	3.6	3.6
10	2000	50	4.0	4.0	4.0

The combined standards can be stored at a temperature of 4 °C for at least one year; the calibration solutions must be freshly prepared when needed.

3 Sample collection and preparation

A filter holder containing a Teflon filter and a silica gel tube are used for sampling. They are connected with silicone tubing so that the dead volume is as small as possible. The filter collects the chlorides, bromides and nitrates ubiquitously present in the air.
With a pump equipped with a flow regulator, air is drawn through the filter/tube combination at a flow rate of 0.25 L/min. After sampling, the filter holder and tube are closed with the plastic caps provided.
Only the silica gel tube is used for the analysis. The sampling and control phases of the silica gel are separated and each transferred to a vessel with a screw top and covered with 10 mL elution solution. The vessels are closed, treated for 15 minutes in an ultrasonic bath and left to stand for 30 minutes.
With a disposable syringe, liquid is taken from the supernatant solution and transferred to an autosampler vessel through a disposable filter.

4 Operating conditions for ion chromatography

Column:	200 × 4.0 mm IonPac AS12A, from Dionex
Precolumn:	50 × 4.0 mm IonPac AS12A, from Dionex
Suppressor:	ASRS (4 mm) with SRS controller, from Dionex
Temperature:	35 °C
Detector:	Conductivity detector
Solvent:	0.0027 M Na_2CO_3, 0.0003 M $NaHCO_3$ in ultrapure water
Pressure:	approx. 1.1 10^5 hPa
Flow rate:	1.5 mL/min
Injection volume:	50 µL

Figure 1 shows an example of a chromatogram obtained under the conditions given above.

5 Analytical determination

A volume of 50 µL of the sample solution prepared as described in Section 3 is injected into the injection loop of the ion chromatograph with a microliter syringe and analysed under the conditions given above.
If the analytical results are not within the calibration range, the samples must be diluted appropriately and analysed again.

6 Calibration

The calibration solutions described in Section 2.4 are used to draw a calibration curve. Volumes of 50 μL of each of the calibration solutions are injected and analysed as for the sample solutions. The peak areas obtained are plotted against the corresponding concentrations. The calibration curves for hydrogen chloride, hydrogen bromide and nitric acid are linear in the range from 0.3 mg/L to 5 mg/L. Figure 2 shows examples of the calibration curves.
To check the calibration curve, a control sample should be analysed each day.
The calibration curve must be prepared again if the analytical conditions change or quality control shows this to be necessary.

7 Calculation of the analytical result

The concentrations of HCl, HBr and HNO_3 in the workplace air are calculated using the concentrations of the substances in the solution calculated by the data processing unit. The data processing unit uses the calibration function calculated from the calibration curve.
The concentrations of the acids in the workplace air are calculated from the concentrations of the solutions, taking into account dilutions and the sampled air volume.
The peak area of the sample signal is calculated according to the following equation:

$$F_{cor} = F - F_{blank}$$

The following equations apply for the concentration of the inorganic acids in the workplace air:

$$\rho = \frac{(F_{cor} - a) \cdot f_{acid} \cdot 0.01\,\mathrm{L}}{b \cdot V_{air}} \cdot \frac{273 + t_g}{273 + t_a}$$

At 20 °C and 1013 hPa:

$$\rho_0 = \rho \, \frac{273 + t_a}{293} \cdot \frac{1013}{p_a}$$

The corresponding concentration by volume σ – independent of the state parameters pressure and temperature – is given by:

$$\sigma = \rho_0 \, \frac{V_M}{MG}$$

$$\sigma = \rho \, \frac{273 + t_a}{293} \cdot \frac{1013}{p_a}$$

For HCl, Hbr and HNO_3 at $t_a = 20\,°C$ and $p_a = 1013$ hPa:

$$\sigma\,(HCl) = \rho \cdot 0.660 \, \frac{mL}{mg}$$

$$\sigma\,(HBr) = \rho \cdot 0.297 \, \frac{mL}{mg}$$

$$\sigma\,(HNO_3) = \rho \cdot 0.382 \, \frac{mL}{mg}$$

where:

F	Peak area from the sample chromatogram
F_{blank}	Peak area for the blank value
F_{cor}	Peak area after correction for the blank value
ρ	Concentration by weight of the inorganic acid in the ambient air as a function of t_a and p_a
ρ_0	Concentration by weight of the inorganic acid in the ambient air at 20 °C and 1013 hPa
a	Intercept of the calibration curve with the ordinate
b	Gradient of the calibration curve in L/mg
f_{acid}	Stoichiometric conversion factor $X^- \rightarrow HX$
0.01 L	Conversion factor for the volume of the measured sample
V_{air}	Sampled air volume in m^3
t_g	Temperature in the gasmeter in °C
t_a	Temperature of the ambient air in °C
p_a	Atmospheric pressure of the ambient air in hPa
σ	Concentration by volume of the inorganic acid in the ambient air in mL/m^3
V_M	Molecular volume of the inorganic acid
MG	Molecular weight of the inorganic acid

Where appropriate when calculating the analytical results, the recovery must be taken into account.

8 Reliability of the method

The characteristics of the method were determined according to the standard DIN EN 482 [4].

8.1 Precision

To calculate the precision of repeated analyses, 10 loaded silica gel tubes were analysed at each of three concentrations. Loading of the silica gel tubes was carried out using a dynamic test gas apparatus.

Table 2. Standard deviation (rel.) *s* and mean variation *u*.

Substance	Concentration mg/m³	Standard deviation (rel.) *s* %	Mean variation *u* %
HCl	1.5	7.3*	16.3
	3.5	2.0*	4.5
	7.0	7.3*	16.2
HBr	3.5	2.4	5.3
	10.0	2.5	5.5
	20.0	3.6	8.0
HNO_3	0.5	2.7	5.9
	2.5	3.0	6.6
	5.0	3.6	8.1

* Influenced by relatively high chloride blank values.

8.2 Recovery rate

The recovery was tested for the whole calibration range. A calibration with calibration solutions (see Section 2.4) and a calibration with loaded silica gel tubes were compared. The recovery for all three inorganic acids was a constant 96–97% in the whole calibration range.

8.3 Quantification limit

To determine the quantification limit, the range of scatter of the blank values of the silica gel tubes was evaluated and a calibration curve was drawn for the lower concentration range [5]. The quantification limit for a sampled air volume of 30 L is 0.1 mg/m^3 for all three acids.

8.4 Sample stability

To determine the sample stability, a series of silica gel tubes ($n = 10$) were loaded using a dynamic test gas apparatus. The loaded silica gel tubes were stored at first for 7 days at room temperature and then at 4 °C in the refrigerator. Two tubes were analysed on the day of loading and after each of 7, 14, 21 and 28 days. No changes in the acid concentrations were detected during this period.

8.5 Sources of error

Ion chromatographic determination is very specific for the anions chloride, bromide and nitrate. Potentially interfering organic acids are separated chromatographically from the anions.
A peak caused by the silica gel can interfere with quantification of the chloride anion. This applies particularly with columns which have been in use for some time, as the retention times become shorter with increasing use. This leads to incomplete separation of interfering and component peaks. In such cases, separation can be improved by reducing the eluent concentration (increasing the retention times).

9 Discussion of the method

During sampling, care must be taken that the ubiquitous salts of the three acids are not also collected. As the analysis is specific for the anions, acids and salts cannot be distinguished. To remove the salts, which only occur as aerosol, from the sampled air, a Teflon filter (pore size 0.45 µm) must be placed in front of the silica gel tube (see Section 3).
The blank value of the silica gel tube can have a considerable effect on the analytical result. Most of the commercially available silica gels can have very high chloride blank values. With "normal" silica gels it is possible to minimise the blank value by carrying out a complicated purification process with ultrapure water. The silica gel tubes of type ORBO 53 (from Supelco) which were used in the development of the method contain already purified silica gel and can be used directly.
It is recommended already purified silica gel be used for the determination of the inorganic acids HCl, HBr and HNO_3. The blank value of all new batches of silica gel must be checked. The ubiquitous presence of chlorides and nitrates also leads to high blank values resulting from chemicals and glassware. The blank values of all chemicals and equipment must be carefully checked.

10 References

[1] *National Institute for Occupational Safety and Health* (NIOSH): NIOSH Manual of Analytical Methods, 4th ed., Method No. 7903, Cincinatti 1994.

[2] *Deutsche Forschungemeinschaft.* List of MAK und BAT Values 2001. Commission for the Investigation of Health Hazards of Chemical Compounds in the Work Area. Report No. 37. WILEY-VCH-Verlagsgesellschaft, Weinheim.

[3] *Bundesministerium für Arbeit und Sozialordnung*: TRGS 900. Grenzwerte in der Luft am Arbeitsplatz – MAK- und TRK-Werte –. In: Technische Regeln und Richtlinien des BMA zur Verordnung über gefährliche Stoffe. Bundesarbeitsblatt 10/1996, S. 106–128, mit Änderungen und Ergänzungen Bundesarbeitsblatt 4/1997.

[4] *Europäisches Komitee für Normung* (CEN): DIN EN 482-Arbeitsplatzatmosphäre-Allgemeine Anforderungen an Verfahren zur Messung von chemischen Arbeitstoffen. Brüssel 1994. Beuth Verlag, Berlin 1994.
[5] *Deutsches Institut für Normung e.V.* (DIN): DIN 32645-Chemische Analytik-Nachweis-, Erfassungs- und Bestimmungsgrenze. Beuth Verlag, Berlin 1994.

Author: *D. Breuer*
Examiners: *R. Hebisch, K. Macho*

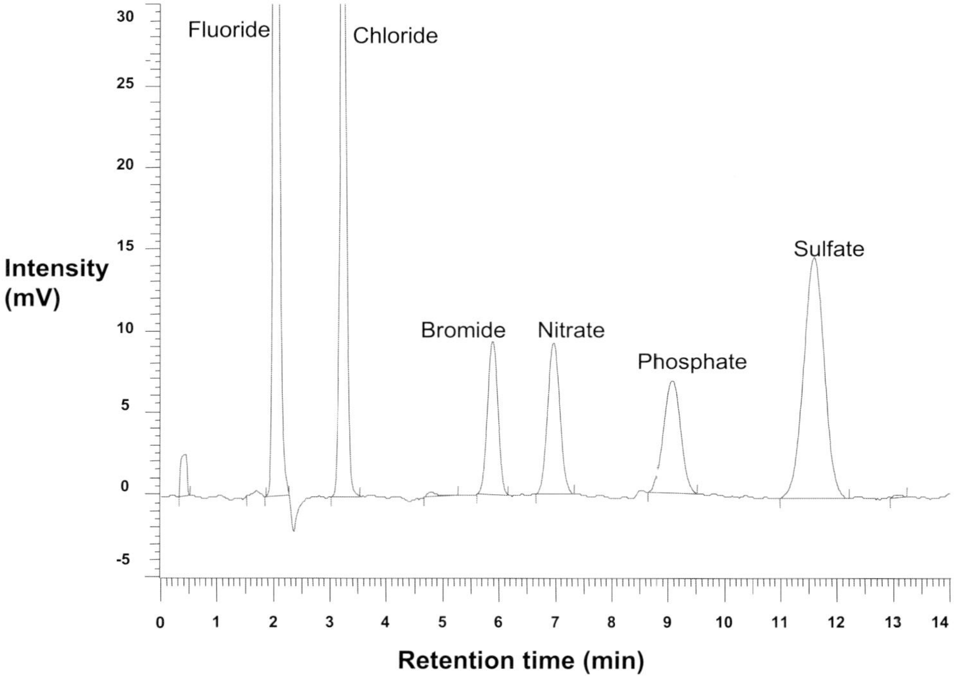

Fig. 1. Example of a chromatogram for the separation of inorganic anions (chromatographic conditions see Section 4). The concentrations are 2.1 mg/L for F^-; Cl^-; Br^-; NO_3^- and 4.2 mg/L for SO_4^{2-}; PO_4^{3-}.

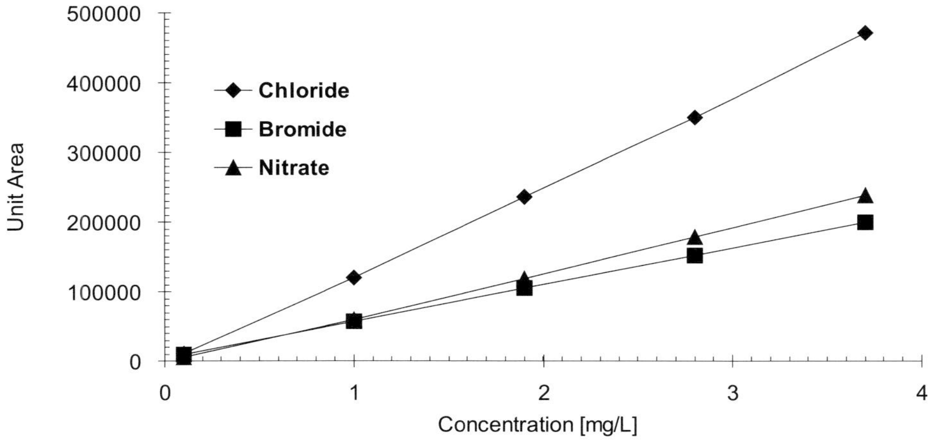

Fig. 2. Calibration curve for inorganic acids (chloride, bromide, nitrate).

Members and Guests of the Working Subgroup

Leader

Prof. Dr. rer. nat. Dr. h. c. A. Kettrup
Institut für Ökologische Chemie
GSF – Forschungszentrum für Umwelt und Gesundheit
Postfach 1129
85758 Neuherberg

Members

Prof. Dr. J. Angerer
Institut für Arbeits-, Sozial- und
Umweltmedizin der Universität Erlangen-Nürnberg
Schillerstraße 25/29
91054 Erlangen

Dr. M. Blaszkewicz
Institut für Arbeitsphysiologie
ZE Analytische Chemie
Ardeystraße 67
44139 Dortmund

Dr. R. Heinrich-Ramm
Zentralinstitut für Arbeitsmedizin
Adolph-Schönfelder-Str. 5
22083 Hamburg

Prof. Dr. W. Riepe
Universität Salzburg
Inst. Chemie u. Biochemie
Hellbrunnerstr. 34
A-5020 Salzburg

Guests

Dr. M. Alker
Infraserv, Höchst
Division ESHAS und Entsorgung
Arbeitsschutz und Anlagensicherheit
Industriepark Höchst/Gebäude C 769
65926 Frankfurt am Main

Dr. M. Ball
Ergo-Forschungsgesellschaft GmbH
Geierstraße 1
22305 Hamburg

Dr. D. Breuer
Berufsgenossenschaftliches Institut
für Arbeitssicherheit-BIA
Alte Heerstr. 111
53754 Sankt Augustin

Dr. D. Franke
Infracor GmBH, IR-AS-U
Gebäude/PB: 9015/10
Paul-Baumannstrasse 1
45764 Marl

Dr. H. Fricke
Bergbau-Berufsgenossenschaft
Hauptverwaltung Hunscheidtstraße 18
44789 Bochum

Dr. H.G. Gielen
Landesamt für Umweltschutz und
Gewerbeaufsicht Rheinland-Pfalz
Rheinallee 97–101
55118 Mainz

Dr. U. Giese
Deutsches Institut für Kautschuktechnologie e.V.
Eupenerstr. 33
30519 Hannover

Dr. C. Habarta
Bayerisches Landesamt für Arbeitsmedizin
und Sicherheitstechnik
Pfarrstraße 3
80538 München

Dr. J.U. Hahn
Berufsgenossenschaftliches Institut
für Arbeitssicherheit-BIA
Alte Heerstr. 111
53754 St. Augustin

Dr. R. Hebisch
Bundesanstalt für Arbeitsschutz und Arbeitsmedizin
Friedrich-Henkel-Weg 1–25
44149 Dortmund

Dr. E. Hellpointner
Bayer AG
PF-Zentrum Monheim, Geb. 6660
51368 Leverkusen

Dipl.-Ing. M. Hennig
Berufsgenossenschaftliches Institut
für Arbeitssicherheit-BIA
Alte Heerstr. 111
53754 St. Augustin

Dr. W. Kleiböhmer
Institut für Chemo- und Biosensorik e.V.
Mendelstr. 7
48149 Münster

Dr. W. Krämer
BASF-AG, Labor für Umweltanalytik
Abt. DUU/OU – Z 570
67056 Ludwigshafen

Dr. M. Kuck
BAYER AG, ZF-ZAL Geb. 013
51368 Leverkusen

Dipl.-Ing. U. Lehnert
Wehrwissenschaftliches Institut
für Werk, Explosiv und Betriebsstoffe (WIWEB)
Landshuterstr. 70
85435 Erding

Dr. N. Lichtenstein
Berufsgenossenschaftliches Institut
für Arbeitssicherheit-BIA
Alte Heerstr. 111
53754 St. Augustin

Dr. C.-P. Maschmeier
Landesamt für Arbeitsschutz des
Landes Sachsen-Anhalt
Kühmauer Str. 70
06846 Dessau

Dr. R. Meyer zu Reckendorf
Zentrallabor, SGL Carbon GmbH
Werner-von-Siemens Str. 18
86405 Meitingen

Dipl.-Ing. K.H. Pannwitz
Drägerwerk AG
Moislinger Allee 53–55
23542 Lübeck

Prof. Dr. Dr. H. Parlar
Lehrstuhl für Chemisch-Technische Analyse
und Chemische Lebensmitteltechnologie
Weihenstephaner Steig 23
85350 Freising-Weihenstephan

Dr. B. Schneider
Clariant GmbH – Division Feinchemikalien
Analytik Intermedites
Stroofstr. 27
65933 Frankfurt/Main

Dr. D. Stevenz
BG Chemie
Gebäude E
Analytisches Labor Leuna
Rudolf-Breitscheid-Straße 18
06237 Leuna

Dr. R. Tribolet
Syngenta AG
R-1094. 1. 78
CH-4002 Basel

Dipl.-Ing. M. Tschickardt
Landesamt für Umweltschutz und Gewerbeaufsicht Rheinland-Pfalz
Rheinallee 97–101
55118 Mainz

Dr. A. Wrede
Aventis Crop Science GmbH
Industriepark Höchst G864
65926 Frankfurt/Main

Secretariat of the commission

Dr. M.R. Lahaniatis
Institut für Ökologische Chemie
GSF-Forschungszentrum für Umwelt und Gesundheit
Postfach 1129
85758 Neuherberg

Dr. R. Schwabe
MAK-Sekretariat
Hochenbacherstr. 15–17
85350 Freising-Weihenstephan

Contents of Volumes 1-6

CAS No.	Substance	Method	Vol.
330-54-1	3-(3,4-dichlorophenyl)-1,1-dimethyl-urea (Diuron)	Urea herbicides	3
330-55-2	3-(3,4-dichlorophenyl)-1-methoxy-1-methylurea (Linuron)	Urea herbicides	3
542-88-1	Bis(chloromethyl)ether	Bis(chloromethyl)ether (BCME)	5
555-37-3	3-(3,4-dichlorophenyl)-1-methoxy-1-*n*-butylurea (Neburon)	Urea herbicides	3
584-84-9	2,4-Toluylene diisocyanate	Hexamethylene diisocyanate (HDI)	1
598-56-1	Dimethylethylamine	Dimethylethylamine, Triethylamine	1
601-77-4	*N*-Nitrosodiisopropylamine	*N*-Nitrosamines	4
602-01-7	1-Methyl-2,3-dinitrobenzene	Dinitrotoluenes	5
606-20-2	1-Methyl-2,6-dinitrobenzene	Dinitrotoluenes	5
610-39-9	1-Methyl-3,4-dinitrobenzene	Dinitrotoluenes	5
612-64-6	*N*-Nitrosoethylphenylamine	N-Nitrosomethylphenylamine (NMPA) and N-nitrosoethylphenylamine (NEPA)	5
614-00-6	N-Nitrosomethylphenylamine	N-Nitrosomethylphenylamine (NMPA) and N-nitrosoethylphenylamine (NEPA)	5
618-85-9	1-Methyl-3,5-dinitrobenzene	Dinitrotoluenes	5
621-64-7	*N*-Nitrosodipropylamine	*N*-Nitrosamines	4
683-18-1	Dibutyltin dichloride	Organotin compounds	3
822-06-0	Hexamethylene diisocyanate	Hexamethylene diisocyanate (HDI)	1
872-50-4	*N*-Methyl-2-pyrrolidone	*N*-Methyl-2-pyrrolidone	1
924-16-3	*N*-Nitrosodibutylamine	*N*-Nitrosamines	4
930-55-2	*N*-Nitrosopyrrolidine	*N*-Nitrosamines	4
1116-54-7	*N*-Nitrosodiethanolamine	*N*-Nitrosodiethanolamine	4
1118-46-3	Butyltin trichloride	Organotin compounds	3
1461-22-9	Tributyltin chloride	Organotin compounds	3
1461-25-2	Tetrabutyltin	Organotin compounds	3
1746-81-2	3-(4-chlorophenyl)-1-methoxy-1-methylurea (Monolinuron)	Urea herbicides	3
2243-62-1	1,5-Diaminonaphthalene	1,5-Diaminonaphthalene	5
3060-89-7	3-(4-bromophenyl)-1-methoxy-1-methylurea (Metobromuron)	Urea herbicides	3
7439-92-1	Lead	Lead	1
7440-02-0	Nickel	Nickel	1
7440-31-5	Tin	Total tin	2
		Organotin compounds	3
7664-38-2	Phosphoric acid	Inorganic acid mists	6
7440-43-9	Cadmium	Cadmium	4
7440-47-3	Chromium	Chromium	1
7440-48-4	Cobalt	Cobalt	1